Applied Calculus of
Variations for Engineers
Third Edition

Applied Calculus of
Variations for Engineers
Third Edition

Louis Komzsik

CRC Press
Taylor & Francis Group
Boca Raton London New York

CRC Press is an imprint of the
Taylor & Francis Group, an **informa** business

CRC Press
Taylor & Francis Group
6000 Broken Sound Parkway NW, Suite 300
Boca Raton, FL 33487-2742

First issued in paperback 2022

© 2020 by Taylor & Francis Group, LLC
CRC Press is an imprint of Taylor & Francis Group, an Informa business

No claim to original U.S. Government works

ISBN-13: 978-0-367-37609-3 (hbk)
ISBN-13: 978-1-03-233757-9 (pbk)
DOI: 10.1201/9781003009740

Library of Congress Cataloging-in-Publication Data

Names: Komzsik, Louis, author.
Title: Applied calculus of variations for engineers / by Louis Komzsik.
Description: Third edition. | Boca Raton, FL : CRC Press/Taylor and Francis, [2020] | Includes bibliographical references and index.
Identifiers: LCCN 2019046081 (print) | LCCN 2019046082 (ebook) | ISBN 9780367376093 (hardback ; acid-free paper) | ISBN 9781003009740 (ebook)
Subjects: LCSH: Calculus of variations. | Engineering mathematics.
Classification: LCC TA347.C3 K66 2020 (print) | LCC TA347.C3 (ebook) | DDC 620.001/51564--dc23
LC record available at https://lccn.loc.gov/2019046081
LC ebook record available at https://lccn.loc.gov/2019046082

Visit the Taylor & Francis Web site at
http://www.taylorandfrancis.com

and the CRC Press Web site at
http://www.crcpress.com

To my daughter, Stella

Contents

Preface

Since the publication of the first edition a dozen years ago, and the second about six years ago, the use of this book has been gradually extended from industry to academia.

Specifically, the first part of the book has been used to teach calculus of variations for undergraduate mathematics students. Furthermore, the second part has been used for mathematical modeling classes for applied mathematics and engineering students.

To accommodate that change of focus and to respond to some requests from students and teachers, this edition brings extensions to numerous sections of both parts. Detailed explanations, illustrative examples, and exercises were added to every chapter of the first part.

Several new sections and subsections modeling various physical phenomena were included in the second part to enhance the mathematical modeling teaching tool set. The new sections and subsections are also indexed and relevant references added.

Hopefully, the students reading this book find the book's theoretical foundation clear and concise, and the analytic and computational examples enlightening. The typographical errors found in the prior editions have been corrected, and a strong effort was made to avoid introducing any in the new material in order to make this work as flawless as possible.

Acknowledgments

I deeply appreciate the contribution of Ms. Lora Dianne Weiss, PhD candidate at UCI and my teaching assistant during the last quarter using the prior edition of this book. Her meticulous review of the new sections, contribution to, and verification of the examples and exercises were invaluable in producing this edition.

My special thanks are due to Nora Konopka, publisher of Taylor & Francis books, and Michele Smith, editor of engineering, for their support of the earlier editions of this book and their appreciation of the value of a new edition with a changed focus. I also thank Michele Dimont, project editor, and Prachi Mishra, editorial assistant, for their work in the preparation of this volume. The author photo is courtesy of Karoly Tihanyi Photography, Budapest.

I am still indebted to the reviewers and contributors of the prior editions: Dr. Leonard Hoffnung of MathWorks, Mr. Chris Mehling of Siemens (retired), Dr. John Brauer of the Milwaukee School of Engineering (retired), and Professor Bajcsay Pál of the Technical University of Budapest (retired).

Dr. Louis Komzsik
University of California, Irvine
September 2019

Author

 **Dr. Louis Komzsik** is a graduate of the Technical University of Budapest, and the Eötvös Lóránd University, both in Hungary. He worked in the industry as an engineering mathematician for 42 years and during those years also lectured as a Visiting Professor at various southern California colleges and universities. Since his retirement, he has been lecturing in the Mathematics Department of the University of California at Irvine.

Dr. Komzsik is the author of a dozen books, among them the widely known *The Lanczos Method* that has also been published in Japanese, Chinese and Hungarian. His *Computational Techniques of Finite Element Analysis* and *Approximation Techniques for Engineers* are both in second editions published by Taylor & Francis. He is also the coauthor of the book *Computational Techniques of Rotor Dynamics with the Finite Element Method.* He owns several patents in mathematical modeling of various engineering phenomena.

Introduction

Calculus of variations has a long history. Its fundamentals were laid down by icons of mathematics like Euler and Lagrange. It was once heralded as the panacea for all engineering optimization problems by suggesting that all one needed to do was to state a variational problem, apply the appropriate Euler-Lagrange equation, and solve the resulting differential equation.

This, as most all encompassing solutions, turned out to be not always true and the resulting differential equations are not necessarily easy to solve. On the other hand, many of the differential equations commonly used by engineers today are derived from a variational problem. Hence, it is important and useful for engineers to delve into this topic.

The book is organized into two parts: calculus of variations foundation, and mathematical modeling of various physical and engineering phenomena. The first part starts with the statement of the fundamental variational problem and its solution via the Euler-Lagrange equation. This is followed by the gradual extension to variational problems subject to constraints, containing functions of multiple variables, and functionals with higher order derivatives. It continues with the inverse problem and analytic solutions, and concludes with approximate solution techniques of variational problems, such as the Ritz, Galerkin, and Kantorovich methods.

With the emphasis on mathematical modeling, the second part starts with a detailed discussion of the geodesic concept of differential geometry and its extensions to higher order spaces. The computational geometry chapter covers the variational origin of natural splines and the variational formulation of B-splines under various constraints.

The chapter dealing with the variational foundation of various motion phenomena will include discussion of orbital motion and it also introduces Hamilton's principle and Lagrange's equations of motion. The penultimate chapter focuses on the variational modeling of several classical mechanical problems. Finally, the fundamental applications of elasticity, heat conduction, and fluid mechanics, are discussed using the computational technology of finite elements.

Part I

Mathematical foundation

1

The foundations of calculus of variations

The topic of the calculus of variations evolves from the analysis of functions. In the analysis of functions the focus is on the relation between two sets of numbers, the independent (x) and the dependent (y) set. The function f creates a one-to-one correspondence between these two sets, denoted as $y = f(x)$.

In this chapter we generalize the concept of functions by allowing the members of the dependent set not being restricted to be numbers, but to be functions themselves. The relationship between these sets is now called a functional. The chapter will describe the fundamental problem of calculus of variations as finding extrema of functionals, most commonly formulated in the form of an integral. The sufficient and necessary conditions for finding the extrema will be presented and several classical problems solved.

1.1 The fundamental problem and lemma of calculus of variations

The fundamental problem of the calculus of variations is to find the extremum (maximum or minimum) of the functional

$$I(y) = \int_{x_0}^{x_1} f(x, y, y')dx,$$

where the solution satisfies the boundary conditions

$$y(x_0) = y_0$$

and

$$y(x_1) = y_1.$$

These problems may also be extended with constraints, the topic of Chapter 2. They may also be generalized to the cases when higher derivatives or multiple functions are given and will be discussed in Chapters 3 and 4, respectively.

A solution process may be arrived at with the following logic. Let us assume that there exists such a solution $y(x)$ for the above problem that satisfies the boundary conditions and produces the extremum of the functional. Furthermore, we assume that it is twice differentiable. In order to prove that this function results in an extremum, we need to prove that any alternative function does not attain the extremum.

We introduce an alternative solution function of the form:

$$Y(x) = y(x) + \epsilon\eta(x),$$

where ϵ is a small undefined number, and $\eta(x)$ is an arbitrary auxiliary function of x, that is also twice differentiable and vanishes at the boundary:

$$\eta(x_0) = \eta(x_1) = 0.$$

In consequence, the following is also true:

$$Y(x_0) = y(x_0) = y_0$$

and

$$Y(x_1) = y(x_1) = y_1.$$

A typical relationship between these functions is shown in Figure 1.1 where the function is represented by the solid line and the alternative function by the dotted line. The dashed line represents the arbitrary auxiliary function.

Since the alternative function $Y(x)$ also satisfies the boundary conditions of the functional, we may substitute into the variational problem.

$$I(\epsilon) = \int_{x_0}^{x_1} f(x, Y, Y')dx$$

where

$$Y'(x) = y'(x) + \epsilon\eta'(x).$$

The new functional is identical with the original in the case when $\epsilon = 0$ and has its extremum when

$$\frac{\partial I(\epsilon)}{\partial \epsilon}\bigg|_{\epsilon=0} = 0.$$

Executing the derivation and taking the derivative into the integral, since the limits are fixed, with the chain rule we obtain

$$\frac{\partial I(\epsilon)}{\partial \epsilon} = \int_{x_0}^{x_1} \left(\frac{\partial f}{\partial Y}\frac{dY}{d\epsilon} + \frac{\partial f}{\partial Y'}\frac{dY'}{d\epsilon}\right)dx.$$

Clearly

$$\frac{dY}{d\epsilon} = \eta(x),$$

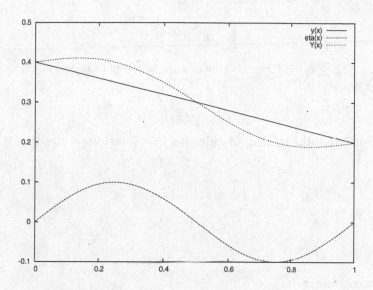

FIGURE 1.1 Alternative solutions example

and

$$\frac{dY'}{d\epsilon} = \eta'(x),$$

resulting in

$$\frac{\partial I(\epsilon)}{\partial \epsilon} = \int_{x_0}^{x_1} \left(\frac{\partial f}{\partial Y} \eta(x) + \frac{\partial f}{\partial Y'} \eta'(x) \right) dx.$$

Integrating the second term by parts yields

$$\int_{x_0}^{x_1} \left(\frac{\partial f}{\partial Y'} \eta'(x) \right) dx = \frac{\partial f}{\partial Y'} \eta(x) \Big|_{x_0}^{x_1} - \int_{x_0}^{x_1} \left(\frac{d}{dx} \frac{\partial f}{\partial Y'} \right) \eta(x) dx.$$

Due to the boundary conditions, the first term vanishes. With substitution and factoring the auxiliary function, the problem becomes

$$\frac{\partial I(\epsilon)}{\partial \epsilon} = \int_{x_0}^{x_1} \left(\frac{\partial f}{\partial Y} - \frac{d}{dx} \frac{\partial f}{\partial Y'} \right) \eta(x) dx.$$

The extremum is achieved when $\epsilon = 0$ as stated above, hence

$$\frac{\partial I(\epsilon)}{\partial \epsilon} \Big|_{\epsilon=0} = \int_{x_0}^{x_1} \left(\frac{\partial f}{\partial y} - \frac{d}{dx} \frac{\partial f}{\partial y'} \right) \eta(x) dx.$$

Let us now consider the following integral:

$$\int_{x_0}^{x_1} \eta(x)F(x)dx,$$

where $x_0 \leq x \leq x_1$ and $F(x)$ is continuous, while $\eta(x)$ is continuously differentiable, satisfying

$$\eta(x_0) = \eta(x_1) = 0.$$

The fundamental lemma of calculus of variations states that if for all such $\eta(x)$

$$\int_{x_0}^{x_1} \eta(x)F(x)dx = 0,$$

then

$$F(x) = 0$$

in the whole interval.

The following proof by contradiction is from [18]. Let us assume that there exists at least one such location $x_0 \leq \zeta \leq x_1$ where $F(x)$ is not zero, for example,

$$F(\zeta) > 0.$$

By the condition of continuity of $F(x)$, there must be a neighborhood of

$$\zeta - h \leq \zeta \leq \zeta + h$$

where $F(x) > 0$. In this case, however, the integral becomes

$$\int_{x_0}^{x_1} \eta(x)F(x)dx > 0,$$

for the right choice of $\eta(x)$, which contradicts the original assumption. Hence, the statement of the lemma must be true.

Applying the lemma to this case results in the **Euler-Lagrange differential equation** specifying the extremum:

$$\frac{\partial f}{\partial y} - \frac{d}{dx}\frac{\partial f}{\partial y'} = 0.$$

Let us immediately illustrate the use of this equation to solve a variational problem posed as

$$I = \int_0^1 \left(\frac{1}{2}y'^2 + (x+1)y\right) dx = \text{extremum}$$

with boundary conditions

$$y(0) = 0, y(1) = 1.$$

The components of the Euler-Lagrange differential equation are

$$\frac{\partial f}{\partial y} = (x+1),$$

$$\frac{d}{dx}\frac{\partial f}{\partial y'} = y''$$

and the differential equation becomes

$$\frac{\partial f}{\partial y} - \frac{d}{dx}\frac{\partial f}{\partial y'} = (x+1) - y'' = 0.$$

Separating the variables we obtain

$$y'' = \frac{d^2y}{dx^2} = x + 1,$$

and integrating produces

$$y' = \frac{x^2}{2} + x + c_1.$$

Integrating again brings the general solution as

$$y = \frac{x^3}{6} + \frac{x^2}{2} + c_1 x + c_2.$$

Using the boundary conditions resolves the integration constants as

$$y(0) = 0 = c_2,$$

and

$$y(1) = 1 = \frac{1^3}{6} + \frac{1^2}{2} + c_1$$

brings

$$c_1 = \frac{1}{3}.$$

Hence, the specific solution is

$$y = \frac{1}{6}x^3 + \frac{1}{2}x^2 + \frac{1}{3}x.$$

Most of the time there is no need to compute the actual extremum, but for this introductory example we will do so. The derivative of the solution is

$$y' = \frac{x^2}{2} + x + \frac{1}{3}.$$

Squaring this, rebuilding the functional and executing the integration produce the actual extremum. In practice, the subject of interest is the solution function and not the actual extremum; hence, it will be seldom computed in the remainder of this book.

1.2   The Legendre test

The Euler-Lagrange differential equation just introduced represents a necessary, but not sufficient, condition for the solution of the fundamental variational problem.

The alternative functional of

$$I(\epsilon) = \int_{x_0}^{x_1} f(x, Y, Y')dx,$$

may be expanded as

$$I(\epsilon) = \int_{x_0}^{x_1} f\left(x, y + \epsilon\eta(x), y' + \epsilon\eta'(x)\right) dx.$$

Assuming that the f function has continuous partial derivatives, the mean-value theorem is applicable:

$$f\left(x, y + \epsilon\eta(x), y' + \epsilon\eta'(x)\right) = f(x, y, y') +$$

$$\epsilon\left(\eta(x)\frac{\partial f(x, y, y')}{\partial y} + \eta'(x)\frac{\partial f(x, y, y')}{\partial y'}\right) + O(\epsilon^2).$$

By substituting we obtain

$$I(\epsilon) = \int_{x_0}^{x_1} f(x, y, y')dx +$$

$$\epsilon\int_{x_0}^{x_1}\left(\eta(x)\frac{\partial f(x, y, y')}{\partial y} + \eta'(x)\frac{\partial f(x, y, y')}{\partial y'}\right) dx + O(\epsilon^2).$$

With the introduction of

$$\delta I_1 = \epsilon\int_{x_0}^{x_1}\left(\eta(x)\frac{\partial f(x, y, y')}{\partial y} + \eta'(x)\frac{\partial f(x, y, y')}{\partial y'}\right) dx,$$

we can write

$$I(\epsilon) = I(0) + \delta I_1 + O(\epsilon^2),$$

where δI_1 is called the first variation. The vanishing of the first variation is a necessary, but not sufficient, condition to have an extremum. To establish a sufficient condition, assuming that the function is three times continuously differentiable, we further expand as

$$I(\epsilon) = I(0) + \delta I_1 + \delta I_2 + O(\epsilon^3).$$

Here the newly introduced second variation is

$$\delta I_2 = \frac{\epsilon^2}{2} \int_{x_0}^{x_1} \left(\eta^2(x) \frac{\partial^2 f(x, y, y')}{\partial y^2} + 2\eta(x)\eta'(x) \frac{\partial^2 f(x, y, y')}{\partial y \partial y'} + \right.$$

$$\left. \eta'^2(x) \frac{\partial^2 f(x, y, y')}{\partial y'^2} \right) dx.$$

We now possess all the components to test for the existence of the extremum (maximum or minimum). The **Legendre test** states that if independently of the choice of the auxiliary $\eta(x)$ function

- the Euler-Lagrange equation is satisfied,

- the first variation vanishes ($\delta I_1 = 0$), and

- the second variation does not vanish ($\delta I_2 \neq 0$)

over the interval of integration, then the functional has an extremum. This test presents the necessary and sufficient conditions for the existence of the extremum. Specifically, the extremum will be a maximum if the second variation is negative, and conversely a minimum if it is positive. Certain similarities to the extremum evaluation of regular functions by the teaching of classical calculus are obvious.

We finally introduce the variation of the function as

$$\delta y = Y(x) - y(x) = \epsilon \eta(x),$$

and the variation of the derivative as

$$\delta y' = Y'(x) - y'(x) = \epsilon \eta'(x).$$

Based on these variations, we distinguish between the following cases:

- strong extremum occurs when δy is small, however, $\delta y'$ is large, while

- weak extremum occurs when both δy and $\delta y'$ are small.

On a final note: the above considerations did not ever state the finding or presence of an absolute extremum; only the local extremum in the interval of the integrand is obtained.

1.3   The Euler–Lagrange differential equation

Let us expand the derivative in the second term of the Euler-Lagrange differential equation as follows:

$$\frac{d}{dx}\frac{\partial f}{\partial y'} = \frac{\partial^2 f}{\partial x \partial y'} + \frac{\partial^2 f}{\partial y \partial y'}y' + \frac{\partial^2 f}{\partial y'^2}y''.$$

This demonstrates that the Euler-Lagrange equation is usually of second order:

$$\frac{\partial f}{\partial y} - \frac{\partial^2 f}{\partial x \partial y'} - \frac{\partial^2 f}{\partial y \partial y'}y' - \frac{\partial^2 f}{\partial y'^2}y'' = 0.$$

The above form is also called the extended form. Consider the case when the multiplier of the second derivative term vanishes:

$$\frac{\partial^2 f}{\partial y'^2} = 0.$$

In this case f must be a linear function of y', in the form of

$$f(x, y, y') = p(x, y) + q(x, y)y'.$$

For this form, the other derivatives of the equation are computed as

$$\frac{\partial f}{\partial y} = \frac{\partial p}{\partial y} + \frac{\partial q}{\partial y}y',$$

$$\frac{\partial f}{\partial y'} = q,$$

$$\frac{\partial^2 f}{\partial x \partial y'} = \frac{\partial q}{\partial x},$$

and

$$\frac{\partial^2 f}{\partial y \partial y'} = \frac{\partial q}{\partial y}.$$

Substituting results in the Euler-Lagrange differential equation of the form

$$\frac{\partial p}{\partial y} - \frac{\partial q}{\partial x} = 0,$$

or

$$\frac{\partial p}{\partial y} = \frac{\partial q}{\partial x}.$$

In order to have a solution, this must be an identity, in which case there must be a function of two variables

$$u(x, y)$$

whose total differential is of the form

$$du = p(x,y)dx + q(x,y)dy = f(x,y,y')dx.$$

The functional may be evaluated as

$$I(y) = \int_{x_0}^{x_1} f(x,y,y')dx = \int_{x_0}^{x_1} du = u(x_1,y_1) - u(x_0,y_0).$$

It follows from this that the necessary and sufficient condition for the solution of the Euler-Lagrange differential equation is that the integrand of the functional be the total differential with respect to x of a certain function of both x and y.

Considering furthermore, that the Euler-Lagrange differential equation is linear with respect to f, it also follows that a term added to f will not change the necessity and sufficiency of that condition.

Another special case may be worthy of consideration. Let us assume that the integrand does not explicitly contain the x term. Then by executing the differentiations we obtain

$$\frac{d}{dx}\left(y'\frac{\partial f}{\partial y'} - f\right) = y'\frac{d}{dx}\frac{\partial f}{\partial y'} - \frac{\partial f}{\partial x} - \frac{\partial f}{\partial y}y' = y'\left(\frac{d}{dx}\frac{\partial f}{\partial y'} - \frac{\partial f}{\partial y}\right) - \frac{\partial f}{\partial x}.$$

With the last term vanishing in this case, the differential equation simplifies to

$$\frac{d}{dx}\left(y'\frac{\partial f}{\partial y'} - f\right) = 0.$$

Its consequence is the expression known as **Beltrami's formula**:

$$y'\frac{\partial f}{\partial y'} - f = c_1, \tag{1.1}$$

where the right-hand side term is an integration constant.

To illustrate the use of Beltrami's solution, we consider

$$I = \int_0^2 y^2(1-y')^2 dx = \text{extremum}$$

with boundary conditions

$$y(0) = 0, y(2) = 1.$$

This is a Beltrami case as the function is independent of x. Since

$$\frac{\partial f}{\partial y'} = -2y^2(1-y'),$$

the formula (multiplied by -1) dictates

$$f - y' \frac{\partial f}{\partial y'} = y^2(1 - y')^2 + 2y'y^2(1 - y') = -c_1.$$

Executing the posted operations and shortening results in

$$y^2 - y^2 y'^2 = -c_1.$$

Algebraic operations reveal the derivative as

$$y' = \frac{\sqrt{y^2 + c_1}}{y}.$$

Separating the variables and integrating yields

$$\sqrt{y^2 + c_1} = x + c_2,$$

from which the implicit solution of

$$y^2 + c_1 = (x + c_2)^2$$

emerges. Applying the boundary conditions

$$y(0) = 0 \rightarrow c_1 = c_2^2,$$

and

$$y(2) = 1 \rightarrow 1 + c_1 = (2 + c_2)^2$$

resolves the integrating coefficients as

$$c_1 = \frac{9}{16}, c_2 = -\frac{3}{4}.$$

Finally the solution is of the form

$$\frac{(x - 3/4)^2}{9/16} - \frac{y^2}{9/16} = 1,$$

which is a hyperbola shifted in the x direction.

The classical problem of the brachistochrone, discussed in the next section, also belongs to this class. It is important to point out that while in most cases when applicable Beltrami's approach is simpler, there are cases when it is not.

Finally, it is also often the case that the integrand does not contain the y term explicitly. Then

$$\frac{\partial f}{\partial y} = 0$$

and the differential equation has the simpler

$$\frac{d}{dx}\frac{\partial f}{\partial y'} = 0$$

form. As above, the result is

$$\frac{\partial f}{\partial y'} = c_2$$

where c_2 is another integration constant. The geodesic problems, subject of Chapter 8, represent this type of Euler-Lagrange equation.

We can surmise that the Euler-Lagrange differential equation's general solution is of the form

$$y = y(x, c_1, c_2),$$

where the c_1, c_2 are constants of integration, and are solved from the boundary conditions

$$y_0 = y(x_0, c_1, c_2)$$

and

$$y_1 = y(x_1, c_1, c_2).$$

1.4　Minimal path problems

This section deals with several classical problems to illustrate the methodology. The problem of finding the minimal path between two points in space will be addressed in different scenarios.

The first problem is simple geometry, the shortest geometric distance between the points. The second one is the well-known classical problem of the brachistochrone, originally posed and solved by Bernoulli. This is the path of the shortest time required to move from one point to the other under the force of gravity. The third problem considers a minimal path in an optical sense and leads to Snell's law of reflection in optics. The fourth example finds the path of minimal kinetic energy of a particle moving under the force of gravity.

All four problems will be presented in two-dimensional space, although they may also be posed and solved in three dimensions with some more algebraic difficulty but without any additional instructional benefit.

1.4.1　Shortest curve between two points

First we consider the rather trivial variational problem of finding the solution of the shortest curve between two points, P_0, P_1, in the plane. The form of the problem using the arc length expression is

$$\int_{P_0}^{P_1} ds = \int_{x_0}^{x_1} \sqrt{1 + y'^2} dx = \text{extremum}.$$

The obvious boundary conditions are the curve going through its endpoints:

$$y(x_0) = y_0,$$

and

$$y(x_1) = y_1.$$

It is common knowledge that the solution in Euclidean geometry is a straight line from point (x_0, y_0) to point (x_1, y_1). The solution function is of the form

$$y(x) = y_0 + m(x - x_0),$$

with slope

$$m = \frac{y_1 - y_0}{x_1 - x_0}.$$

To evaluate the integral, we compute the derivative as

$$y' = m$$

and the function becomes

$$f(x, y, y') = \sqrt{1 + m^2}.$$

Since the integrand is constant, the integral is trivial

$$I(y) = \sqrt{1 + m^2} \int_{x_0}^{x_1} dx = \sqrt{1 + m^2}(x_1 - x_0).$$

The square of the functional is

$$I^2(y) = (1 + m^2)(x_1 - x_0)^2 = (x_1 - x_0)^2 + (y_1 - y_0)^2.$$

This is the square of the distance between the two points in plane; hence, the extremum is the distance between the two points along the straight line. Despite the simplicity of the example, the connection of a geometric problem to a variational formulation of a functional is clearly visible. This will be the most powerful justification for the use of this technique.

Let us now solve the

$$\int_{x_0}^{x_1} \sqrt{1 + y'^2} dx = \text{extremum}$$

problem via its Euler-Lagrange equation form. Note that the form of the integrand lends itself to the use of one of three methods: the standard form, the Beltrami formula, or the extended form. Let us use the extended form.

$$\frac{\partial f}{\partial y} = 0,$$

$$\frac{\partial^2 f}{\partial x \partial y'} = 0,$$

$$\frac{\partial^2 f}{\partial y \partial y'} = 0,$$

and

$$\frac{\partial^2 f}{\partial y'^2} = \frac{1}{(1 + y'^2)^{3/2}}.$$

Substituting into the extended form gives

$$\frac{1}{(1 + y'^2)^{3/2}} y'' = 0,$$

which simplifies into

$$y'' = 0.$$

Integrating twice, one obtains

$$y(x) = c_0 + c_1 x,$$

clearly the equation of a line. Substituting into the boundary conditions, we obtain two equations,

$$y_0 = c_0 + c_1 x_0,$$

and

$$y_1 = c_0 + c_1 x_1.$$

The solution of the resulting linear system of equations is

$$c_0 = y_0 - c_1 x_0,$$

and

$$c_1 = \frac{y_1 - y_0}{x_1 - x_0}.$$

It is easy to reconcile that

$$y(x) = y_0 - \frac{y_1 - y_0}{x_1 - x_0} x_0 + \frac{y_1 - y_0}{x_1 - x_0} x$$

is identical to

$$y(x) = y_0 + m(x - x_0).$$

The noticeable difference between the two solutions of this problem is that using the Euler-Lagrange equation required no a priori assumption on the shape of the curve and the geometric know-how was not used. This is the case in most practical engineering applications and this is the reason for the utmost importance of the Euler-Lagrange equation.

1.4.2   The brachistochrone problem

The problem of the brachistochrone may be the first problem of variational calculus, already solved by Johann Bernoulli in the late 1600s. The name stands for the shortest time in Greek, indicating the origin of the problem.

The problem is elementary in a physical sense. Its goal is to find the shortest path of a particle moving in a vertical plane from a higher point to a lower point under only the force of gravity. The sought solution is the function $y(x)$ with boundary conditions $y(x_0) = y_0$ and $y(x_1) = y_1$ where

$$P_0 = (x_0, y_0)$$

and

$$P_1 = (x_1, y_1)$$

are the starting and terminal points, respectively. Based on elementary physics considerations, the problem represents an exchange of potential energy with kinetic energy.

In a gravitational field, a moving body's kinetic energy is related to its velocity and its mass, its potential energy to the height of its position and its mass. The higher the velocity and the mass, the bigger the kinetic energy. A body can gain kinetic energy using its potential energy, and conversely, can use its kinetic energy to build up potential energy. At any point during the movement, the gain of the kinetic energy is the same as the loss of the potential energy, hence the total energy is at equilibrium.

The potential energy loss of the particle at any x, y point during the motion is

$$\Delta E_p = mg(y_0 - y),$$

where m is the mass of the particle and g is the acceleration of gravity. The kinetic energy gain is

$$\Delta E_k = \frac{1}{2}mv^2$$

assuming that the particle at the (x, y) point has velocity v. They are in balance as

$$\Delta E_k = \Delta E_p,$$

resulting in an expression of the velocity as

$$v = \sqrt{2g(y_0 - y)}.$$

The velocity by definition is

$$v = \frac{ds}{dt},$$

where s is the arc length of the yet unknown curve. The time required to run the length of the curve is

$$t = \int_{P_0}^{P_1} dt = \int_{P_0}^{P_1} \frac{1}{v} ds.$$

Using the arc length formula from calculus, we get

$$t = \int_{x_0}^{x_1} \frac{\sqrt{1 + y'^2}}{v} dx.$$

Substituting the velocity expression yields

$$t = \frac{1}{\sqrt{2g}} \int_{x_0}^{x_1} \frac{\sqrt{1 + y'^2}}{\sqrt{y_0 - y}} dx.$$

Since we are looking for the minimal time, this is a variational problem of

$$I(y) = \frac{1}{\sqrt{2g}} \int_{x_0}^{x_1} \frac{\sqrt{1 + y'^2}}{\sqrt{y_0 - y}} dx = \text{extremum}.$$

The integrand does not contain the independent variable; hence, we can apply Beltrami's formula of Equation (1.1). This results in the form of

$$\frac{y'^2}{\sqrt{(y_0 - y)(1 + y'^2)}} - \frac{\sqrt{1 + y'^2}}{\sqrt{y_0 - y}} = c_0.$$

Creating a common denominator on the left-hand side produces

$$\frac{y'^2 \sqrt{y_0 - y} - \sqrt{1 + y'^2} \sqrt{(y_0 - y)(1 + y'^2)}}{\sqrt{(y_0 - y)(1 + y'^2)} \sqrt{y_0 - y}} = c_0.$$

Grouping the numerator simplifies to

$$\frac{-1}{\sqrt{(y_0 - y)(1 + y'^2)}} = c_0.$$

Canceling and squaring results in the solution for y' as

$$y'^2 = \frac{1 - c_0^2(y_0 - y)}{c_0^2(y_0 - y)}.$$

Since

$$y'^2 = \left(\frac{dy}{dx}\right)^2,$$

the differential equation may be separated as

$$dx = \frac{\sqrt{y_0 - y}}{\sqrt{2c_1 - (y_0 - y)}} dy.$$

Here the new constant is introduced for simplicity as

$$c_1 = \frac{1}{2c_0^2}.$$

Finally x may be expressed directly by integrating

$$x = \int \frac{\sqrt{y_0 - y}}{\sqrt{2c_1 - (y_0 - y)}} dy + c_2.$$

The trigonometric substitution of

$$y_0 - y = 2c_1 \sin^2\left(\frac{t}{2}\right)$$

yields the integral of

$$x = 2c_1 \int \sin^2\left(\frac{t}{2}\right) dt = c_1 \left(t - \sin(t)\right) + c_2,$$

The constants of integration may be solved by substituting the boundary points. At $t = 0$ we easily find

$$x = c_2 = x_0.$$

Reorganizing and some trigonometry yields

$$y = y_0 - c_1 \left(1 - \cos(t)\right).$$

Substituting the endpoint location into the y equation, we obtain

$$y = y_0 - c_1 \left(1 - \cos(t)\right) = y_1,$$

which is inconclusive, since the time of reaching the endpoint is not known. For a simple conclusion of this discussion, let us assume that the particle reaches the endpoint at time $t = \Pi/2$. Then

$$c_1 = y_0 - y_1,$$

and the final solution is

$$x = x_0 + (y_0 - y_1)(t - \sin(t))$$

and
$$y = y_1 + (y_0 - y_1)(1 - \cos(t)).$$

The final solution of the brachistochrone problem therefore is a cycloid.

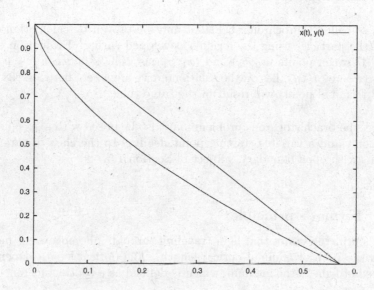

FIGURE 1.2 Solution of the brachistochrone problem

Figure 1.2 depicts the problem of the point moving from $(0, 1)$ until it reaches the x axis.

For the case shown in Figure 1.2, the point moving from $(0, 1)$ until it reaches the x axis, the solution curve is

$$x = t - \sin(t)$$

and

$$y = \cos(t).$$

The resulting curve seems somewhat counter-intuitive, especially in view of the earlier example of the shortest geometric distance between two points in the plane and demonstrated by the straight line chord between the two points. The shortest time, however, when the speed obtained during the traversal of the interval depends on the path taken, is an entirely different matter.

Another intriguing characteristic of the brachistochrone problem is that when two particles are let go from two different points of the curve they will reach the terminal point of the curve at the same time. This is also counter-intuitive, since clearly they have different geometric distances to cover; how-ever, since they are acting under the gravity and the slope of the curve is different at the two locations, the particle starting from a higher location gathers much bigger speed than the particle starting at a lower location.

This, so-called tautochrone, behavior may be proven by calculation of the time of the particles using the formula developed earlier. Evaluation of this integral between points (x_0, y_0) and (x_1, y_1) as well as between (x_2, y_2) and (x_1, y_1) (where (x_2, y_2) lies on the solution curve anywhere between the start-ing and terminal point) will result in the same time.

Hence, the brachistochrone problem may also be posed with a specified ter-minal point and a variable starting point, leading to the class of variational problems with open boundary, subject of Section 1.5.

1.4.3   Fermat's principle

Fermat's principle states that light traveling through inhomogeneous medium chooses the path of minimal optical length. The optical length depends on the speed of light in the medium, which is defined as a continuous function of

$$c(y),$$

where y is the vertical component of the path. Then it may be defined as

$$c(y) = \frac{ds}{dt},$$

the derivative of the length of the path. Similarly to the brachistochrone problem, then the time required to cover the distance between two points is

$$t = \int \frac{1}{c(y)} ds.$$

The problem is now posed as a variational problem of

$$I(y) = \int_{(x_1, y_1)}^{(x_2, y_2)} \frac{ds}{c(y)}.$$

Substituting the arc length results in

$$\int_{x_1}^{x_2} \frac{\sqrt{1 + y'^2}}{c(y)} dx = \text{extremum},$$

with boundary conditions given at the two points P_1, P_2.

$$y(x_1) = y_1; y(x_2) = y_2.$$

The functional does not contain the x term explicitly, allowing the use of Beltrami's formula resulting in the simplified form of

$$y' \frac{\partial f}{\partial y'} - f = k_1$$

where k_1 is a constant of integration and its notation is chosen to distinguish from the speed of light value c. Substituting f, differentiating and simplifying yields

$$\frac{1}{c(y)\sqrt{1 + y'^2}} = -k_1.$$

Reordering and separating results

$$\int dx = \pm k_1 \int \frac{c(y)}{\sqrt{1 - k_1^2 c^2(y)}} dy.$$

Depending on the particular model of the speed of light in the medium, the result varies. In the case of the inhomogeneous optical medium consisting of two homogeneous media in which the speed of light is piecewise constant, the result is the well-known Snell's law describing the scenario of the breaking path of light at the water's surface.

Assume the speed of light is c_1 between points P_1 and P_0 and c_2 between points P_0 and P_2, both constant in their respective medium. The boundary point between the two media is represented by

$$P_0(x, y_0),$$

where the notation signifies the fact that the x location of the light ray is not known yet. The known y_0 location specifies the distance of the points in the two separate media from the boundary.

Then the time to run the full path between P_1 and P_2 is simply

$$t = \frac{\sqrt{(x - x_1)^2 + (y_0 - y_1)^2}}{c_1} + \frac{\sqrt{(x_2 - x)^2 + (y_2 - y_0)^2}}{c_2}.$$

The minimum of this is simply obtained by classical calculus as

$$\frac{dt}{dx} = 0,$$

or

$$\frac{x - x_1}{c_1 \sqrt{(x - x_1)^2 + (y_0 - y_1)^2}} - \frac{x_2 - x}{c_2 \sqrt{(x_2 - x)^2 + (y_2 - y_0)^2}} = 0.$$

The solution of this equation yields the x location of the ray crossing the boundary, and produces the well-known Snell's law of

$$\frac{\sin \phi_1}{c_1} = \frac{\sin \phi_2}{c_2},$$

where the angles are measured with respect to the normal of the boundary between the two media. The preceding work generalizes to multiples of homogeneous media, which is a practical application in lens systems of optical machinery.

1.4.4   Particle moving in a gravitational field

The motion of a particle moving in a gravitational field is computed based on the minimum momentum principle first described by Euler. The principle states that a particle under the influence of a gravitational field moves on a path where

$$I = \int_{P_0}^{P_1} mv\,ds = \text{minimum}.$$

Here m is the mass of the particle, v is the velocity of the particle and their product is the particle's momentum. The boundary points are

$$P_0 = (x_0, y_0), P_1 = (x_1, y_1).$$

Let us assume for the simplicity of algebra, but without the loss of generality of the discussion, that the starting position is at the origin:

$$(x_0, y_0) = (0, 0).$$

Substituting

$$ds = \sqrt{1 + (y')^2}dx,$$

the functional may be written as

$$I = m \int_{x_0}^{x_1} v\sqrt{1 + (y')^2}dx.$$

The potential energy change of the particle at a certain height of y is

$$\Delta E_p = mg(y - y_0) = mgy.$$

The kinetic energy change corresponding to that position is

$$\Delta E_k = \frac{1}{2}m(u^2 - v^2),$$

where u is an initial speed at (x_0, y_0) with yet undefined direction but given magnitude. They are equivalent in the sense that any potential energy gain equals the kinetic energy loss and vice versa. Hence

$$mgy = \frac{1}{2}m(u^2 - v^2),$$

or

$$2gy = u^2 - v^2,$$

from which the velocity at any point of the trajectory is

$$v = \sqrt{u^2 - 2gy}.$$

Substituting into the functional yields

$$I = m \int_{x_0}^{x_1} \sqrt{u^2 - 2gy}\sqrt{1 + (y')^2}dx = \text{extremum}.$$

In the following we assume unit mass for simplicity, $m = 1$. Since the functional does not contain x explicitly, we can use Beltrami's formula. Its derivative term is

$$f_{y'} = \sqrt{u^2 - 2gy}\frac{1}{2}\frac{2y'}{\sqrt{1 + (y')^2}}.$$

Canceling and substituting into the formula yield

$$\sqrt{u^2 - 2gy}\sqrt{1 + (y')^2} - \sqrt{u^2 - 2gy}\frac{y'^2}{\sqrt{1 + (y')^2}} - c_1.$$

Gathering like terms and using a common denominator produce

$$\sqrt{u^2 - 2gy}\frac{((1 + (y')^2) - y'^2)}{\sqrt{1 + (y')^2}} = c_1,$$

or

$$\frac{\sqrt{u^2 - 2gy}}{\sqrt{1 + y'^2}} = c_1,$$

where c_1 is an arbitrary constant. Then

$$u^2 - 2gy = c_1^2(1 + y'^2),$$

and

$$y' = \frac{\sqrt{u^2 - 2gy - c_1^2}}{c_1}$$

Separating as

$$dx = c_1\frac{dy}{\sqrt{u^2 - 2gy - c_1^2}}$$

and integrating yields

$$x - c_2 = c_1\frac{1}{-g}\sqrt{u^2 - 2gy - c_1^2}$$

with c_2 being another constant of integration whose sign is chosen for later convenience. Multiplying by g and squaring results in

$$c_1^2(u^2 - 2gy - c_1^2) = g^2(x - c_2)^2,$$

Reordering yields

$$c_1^2(u^2 - c_1^2) - g^2(x - c_2)^2 = c_1^2 2gy$$

from which the particle trajectory becomes

$$y = \frac{u^2 - c_1^2}{2g} - \frac{g}{2c_1^2}(x - c_2)^2.$$

The resolution of the constants may be accomplished by using the initial location and velocity of the particle. Let us assume $x_0 = 0$ and x_1 is open. The constant c_1 is related to the velocity and constant c_2 is related to the location.

The derivative of the trajectory is

$$y' = \frac{-2g}{2c_1^2}(x - c_2) = -\frac{g}{c_1^2}(x - c_2).$$

Given the initial velocity angle to the x axis as α then

$$y'(0) = \frac{g}{c_1^2}c_2 = \tan(\alpha).$$

This enables expressing

$$c_2 = \frac{c_1^2 \tan(\alpha)}{g}.$$

Let us now substitute into the initial position of the particle

$$y(0) = \frac{u^2 - c_1^2}{2g} - \frac{g}{2c_1^2}(c_2)^2 = \frac{u^2 - c_1^2}{2g} - \frac{g}{2c_1^2}\frac{c_1^4 \tan^2(\alpha)}{g^2} = 0.$$

Then

$$\frac{u^2 - c_1^2}{2g} - \frac{c_1^2 \tan^2(\alpha)}{2g} = 0.$$

This resolves the second constant as

$$u^2 = c_1^2(\tan^2(\alpha) + 1).$$

Substituting into the location function

$$y(x) = \frac{u^2 - c_1^2}{2g} - \frac{g}{2c_1^2}(x - c_2)^2 =$$

$$= \frac{u^2 - c_1^2}{2g} - \frac{g}{2c_1^2}\left(x - \frac{c_1^2 \tan(\alpha)}{g}\right)^2.$$

Executing the square operation brings

$$y(x) = \frac{u^2 - c_1^2}{2g} - \frac{g}{2c_1^2}\left(x^2 - 2x\frac{c_1^2\tan(\alpha)}{g} + \frac{c_1^4\tan^2(\alpha)}{g^2}\right)$$

and multiplying yields

$$y(x) = \frac{u^2 - c_1^2}{2g} - \frac{gx^2}{2c_1^2} + x\tan(\alpha) - \frac{c_1^2\tan^2(\alpha)}{2g}.$$

Finally the expression for the u obtained above and substituted into the location yields

$$y(x) = \frac{c_1^2(\tan^2(\alpha) + 1) - c_1^2}{2g} - \frac{gx^2}{2c_1^2} + x\tan(\alpha) - \frac{c_1^2\tan^2(\alpha)}{2g}$$

and

$$y(x) = \frac{c_1^2\tan^2(\alpha)}{2g} - \frac{gx^2}{2c_1^2} + x\tan(\alpha) - \frac{c_1^2\tan^2(\alpha)}{2g}$$

The first and last terms cancel out resulting in

$$y(x) = x\tan(\alpha) - \frac{gx^2}{2c_1^2}.$$

The second term still contains the c_1 term. Its substitution

$$y(x) = x\tan(\alpha) - \frac{gx^2}{2\frac{u^2}{\tan^2(\alpha)+1}},$$

and some trigonometry

$$\frac{1}{\tan^2(\alpha) + 1} = \frac{1}{\frac{\sin^2(\alpha)}{\cos^2(\alpha)} + 1} = \frac{\cos^2(\alpha)}{\sin^2(\alpha) + \cos^2(\alpha)}$$

yields the explicit trajectory formula for the particle as

$$y = x\tan(\alpha) - \frac{gx^2}{2u^2\cos^2(\alpha)}.$$

This may be reconciled with the well-known parametric trajectory formula from physics that states

$$x(t) = u\cos(\alpha)t$$

and

$$y(t) = u\sin(\alpha)t - \frac{1}{2}gt^2.$$

Expressing the time variable t from the first equation we obtain

$$t = \frac{x(t)}{u\cos(\alpha)}$$

and squaring we obtain

$$t^2 = \frac{x^2}{u^2 \cos^2(\alpha)},$$

which is in agreement with the second part of the explicit trajectory developed above. The first part similarly agrees as

$$y = x \tan(\alpha) = u \cos(\alpha)t \cdot \tan(\alpha) = u \sin(\alpha)t.$$

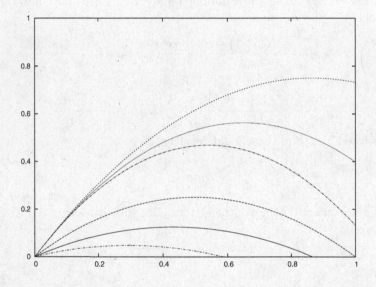

FIGURE 1.3 Trajectory of particle

Figure 1.3 demonstrates the path of the particle. The upper three curves show the path with a 60-degree angle of the initial velocity and with different magnitudes. The lower three curves demonstrate the paths obtained by the same magnitude (10 units), but different angles of the initial velocity. For visualization purposes the gravity constant was chosen to be 10 units as well.

A similarity between the four problems of this section is apparent. This recognition is a very powerful aspect of variational calculus. There are many instances in engineering applications when one physical problem may be solved in an analogous form using another principle. The common variational formulation of both problems is the key to such recognition in most cases.

1.5   Open boundary variational problems

Let us consider the variational problem of Section 1.1:

$$I(y) = \int_{x_0}^{x_1} f(x, y, y')dx,$$

with boundary condition

$$y(x_0) = y_0.$$

Let the boundary condition at the upper end be open. We introduce an auxiliary function $\eta(x)$ that in this case only satisfies the lower boundary condition,

$$\eta(x_0) = 0.$$

The extremum in this case is obtained from the same concept as earlier:

$$\frac{\partial I(\epsilon)}{\partial \epsilon}\bigg|_{\epsilon=0} = \int_{x_0}^{x_1} \left(\frac{\partial f}{\partial y}\eta(x) + \frac{d}{dx}\frac{\partial f}{\partial y'}\eta'(x) \right) dx = 0,$$

while recognizing the fact that x_1 is undefined. Integrating by parts and considering the one-sided boundary condition posed on the auxiliary function yields

$$\frac{\partial I(\epsilon)}{\partial \epsilon}\bigg|_{\epsilon=0} = \frac{\partial f}{\partial y'}\bigg|_{x=x_1}\eta(x_1) + \int_{x_0}^{x_1} \left(\left(\frac{\partial f}{\partial y} - \frac{d}{dx}\frac{\partial f}{\partial y'} \right)\eta'(x) \right) dx = 0.$$

The extremum is obtained when the Euler-Lagrange equation of

$$\frac{\partial f}{\partial y} - \frac{d}{dx}\frac{\partial f}{\partial y'} = 0$$

along with the given boundary condition of

$$y(x_0) = y_0$$

is satisfied, in addition to obeying the auxiliary constraint due to the open boundary condition

$$\frac{\partial f}{\partial y'}\bigg|_{x=x_1} = 0.$$

Similar argument may be applied when the starting point is, or both of them are open. This is a special case of the transversality condition described in more detail in the next chapter in connection with the more generic constrained variational problems.

We illustrate the open boundary scenario with the problem

$$I = \int_0^1 \left(\frac{1}{2} y'^2 + yy' + y' + y \right) dx = \text{extremum}$$

with the boundary conditions open on both locations. The first requirement for the solution is the satisfaction of the Euler-Lagrange differential equation. Its components are

$$\frac{\partial f}{\partial y} = y' + 1,$$

$$\frac{\partial f}{\partial y'} = y' + y + 1,$$

and the equation becomes

$$\frac{\partial f}{\partial y} - \frac{d}{dx} \frac{\partial f}{\partial y'} = y' + 1 - \frac{d}{dx}(y' + y + 1) = 0.$$

This results in the ordinary differential equation of

$$y'' = 1.$$

Note, that since this is a case when Beltrami's solution may also be used; however, in the case of this example it leads to

$$\frac{1}{2} y'^2 - yy' - y = c_1,$$

which is a more difficult solution avenue. By double integration of the first form, the general solution is

$$y = ax + b + \frac{1}{2} x^2.$$

The integrating coefficients are found from the transversality conditions. The left-hand boundary produces

$$\left. \frac{\partial f}{\partial y'} \right|_{x=0} = y'(0) + y(0) + 1 = 0$$

or

$$a + b + 1 = 0,$$

while the right-hand boundary gives

$$\left. \frac{\partial f}{\partial y'} \right|_{x=1} = y'(1) + y(1) + 1 = 0$$

or

$$2a + b + \frac{5}{2} = 0.$$

The solution of the two equations yields

$$a = -\frac{3}{2}$$

and

$$b = \frac{1}{2}.$$

Hence, the solution of the open boundary variational problem is

$$y(x) = \frac{1}{2}x^2 - \frac{3}{2}x + \frac{1}{2}.$$

Computing the derivative of the solution, substitution into the functional and integration yield the extremum if needed.

1.6 Exercises

Find the solutions for the problems by using the Euler-Lagrange differential equation and the boundary conditions, if given.

1.
$I = \int (y'^2 + 2y)dx = $ extremum.

2.
$I = \int (y'^2 + 4xy')dx = $ extremum.

3.
$I = \int (y'^2 + yy' + y^2)dx = $ extremum.

4.
$I = \int (xy'^2 - yy' + y)dx = $ extremum.

5.
$I = \int_1^2 \frac{\sqrt{1+y'^2}}{x}dx = $ extremum.
Boundary conditions: $y(1) = 0, y(2) = 1$.

6.
$I = \int_0^{\frac{\pi}{8}} (y'^2 + 2yy' - 16y^2)dx = $ extremum.
Boundary conditions: $y(0) = 0, y(\pi/8) = 1$.

7.
$I = \int_0^1 (1+x)y'^2 dx = \text{extremum.}$
Boundary conditions: $y(0) = 0, y(1) = 1$.

8.
$I = \int \frac{y'^2}{x^3} dx = \text{extremum.}$

9.
$I = \int (y^2 + y'^2 + 2ye^x) dx = \text{extremum.}$

10.
$I = \int (2y + y'^2) dx = \text{extremum.}$

2

Constrained variational problems

The boundary conditions applied to a variational problem may also be considered as constraints. The subject of this chapter is to generalize the constraint concept in two senses. The first is to allow more difficult, algebraic boundary conditions, and the second is to allow constraints imposed on the interior of the domain as well. Several isoperimetric problems from geometry will be solved, and a closed loop integral solution also presented and illustrated.

2.1 Algebraic boundary conditions

There is the possibility of defining the boundary condition at one end of the integral of the variational problem with an algebraic constraint. Let the

$$\int_{x_0}^{x_1} f(x, y, y')dx = \text{extremum}$$

variational problem be subject to the customary boundary condition

$$y(x_0) = y_0$$

on the lower end, and on the upper end subject of an algebraic condition of the following form:

$$g(x, y) = 0.$$

We again consider an alternative solution of the form

$$Y(x) = y(x) + \epsilon\eta(x).$$

The given boundary condition in this case is

$$\eta(x_0) = 0.$$

Then, following [6], the intersection of the alternative solution and the algebraic curve is

$$X_1 = X_1(\epsilon)$$

and
$$Y_1 = Y_1(\epsilon).$$

The notation is to distinguish from the fixed boundary condition values given via x_1, y_1. Therefore, the algebraic condition is

$$g(X_1, Y_1) = 0.$$

This must be true for any ϵ, hence applying the chain rule yields

$$\frac{dg}{d\epsilon} = \frac{\partial g}{\partial X_1}\frac{dX_1}{d\epsilon} + \frac{\partial g}{\partial Y_1}\frac{dY_1}{d\epsilon} = 0. \tag{2.1}$$

Since

$$Y_1 = y(X_1) + \epsilon\eta(X_1),$$

we expand the last derivative of the second term of Equation (2.1) as

$$\frac{dY_1}{d\epsilon} = \frac{dy}{dx}\Big|_{x=X_1}\frac{dX_1}{d\epsilon} + \eta(X_1) + \epsilon\frac{d\eta}{dx}\Big|_{x=X_1}\frac{dX_1}{d\epsilon}.$$

Substituting into Equation (2.1) results in

$$\frac{dg}{d\epsilon} = \frac{\partial g}{\partial X_1}\frac{dX_1}{d\epsilon} + \frac{\partial g}{\partial Y_1}\Big(\frac{dy}{dx}\Big|_{x=X_1}\frac{dX_1}{d\epsilon} + \eta(X_1) + \epsilon\frac{d\eta}{dx}\Big|_{x=X_1}\frac{dX_1}{d\epsilon}\Big) = 0.$$

Since (X_1, Y_1) becomes (x_1, y_1) when $\epsilon = 0$,

$$\frac{dX_1}{d\epsilon}\Big|_{\epsilon=0} = -\frac{\eta(x_1)\frac{\partial g}{\partial y}\Big|_{y=y_1}}{\frac{\partial g}{\partial x}\Big|_{x=x_1} + \frac{\partial g}{\partial y}\Big|_{y=y_1}\frac{dy}{dx}\Big|_{x=x_1}}. \tag{2.2}$$

We now consider the variational problem of

$$I(\epsilon) = \int_{x_0}^{X_1} f(x, Y, Y')dx.$$

The derivative of this is

$$\frac{\partial I(\epsilon)}{\partial \epsilon} = \frac{dX_1}{d\epsilon}f\Big|_{x=X_1} + \int_{x_0}^{X_1}\Big(\frac{\partial f}{\partial Y}\eta + \frac{\partial f}{\partial Y'}\eta'\Big)dx.$$

Integrating by parts and taking $\epsilon = 0$ yields

$$\frac{\partial I(\epsilon)}{\partial \epsilon}\Big|_{\epsilon=0} = \frac{dX_1}{d\epsilon}\Big|_{\epsilon=0}f\Big|_{x=x_1} + \frac{\partial f}{\partial y'}\Big|_{x=x_1}\eta(x_1) + \int_{x_0}^{x_1}\Big(\frac{\partial f}{\partial y} - \frac{d}{dx}\frac{\partial f}{\partial y'}\Big)\eta dx.$$

Substituting the first expression with Equation (2.2) results in

$$\left(\frac{\partial f}{\partial y'}\Big|_{x=x_1} - \frac{\frac{\partial g}{\partial y}\Big|_{y=y_1}f\Big|_{x=x_1}}{\frac{\partial g}{\partial x}\Big|_{x=x_1} + \frac{\partial g}{\partial y}\Big|_{y=y_1}\frac{dy}{dx}\Big|_{x=x_1}}\right)\eta(x_1) +$$

$$\int_{x_0}^{x_1} \left(\frac{\partial f}{\partial y} - \frac{d}{dx} \frac{\partial f}{\partial y'} \right) \eta dx = 0.$$

Due to the fundamental lemma of calculus of variations, to find the constrained variational problem's extremum, the following conditions need to be satisfied. The Euler-Lagrange differential equation

$$\frac{\partial f}{\partial y} - \frac{d}{dx} \frac{\partial f}{\partial y'} = 0,$$

the given boundary condition

$$y(x_0) = y_0,$$

and the transversality condition of the form

$$\frac{\partial f}{\partial y'} \bigg|_{x=x_1} = \frac{\frac{\partial g}{\partial y} \bigg|_{y=y_1} f \bigg|_{x=x_1}}{\frac{\partial g}{\partial x} \bigg|_{x=x_1} + \frac{\partial g}{\partial y} \bigg|_{y=y_1} \frac{dy}{dx} \bigg|_{x=x_1}}.$$

The transversality condition is named such as it assures that the solution transverses the boundary curve [16]. The evaluation of the transversality condition requires special care when the constraint curve produces infinite derivatives.

2.1.1 Transversality condition computation

To demonstrate the computation of the transversality condition, we consider the problem

$$I = \int_0^{x_1} \frac{\sqrt{1 + y'^2}}{y} dx = \text{extremum}$$

with left-hand side boundary condition

$$y(0) = 0$$

and the right-hand side boundary constraint of

$$y(x_1) = y_1$$

located on the line

$$g(x, y) = y - x + 5 = 0.$$

The first part of the solution as indicated above is to satisfy the Euler-Lagrange differential equation. Its components are

$$\frac{\partial f}{\partial y} = f_y = -\frac{\sqrt{1 + y'^2}}{y^2}$$

and

$$\frac{\partial f}{\partial y'} = f_{y'} = \frac{y'}{y\sqrt{1+y'^2}},$$

with the introduction of a simplified notation for convenience in this section. Then

$$\frac{d}{dx} f_{y'} = \frac{y''y\sqrt{1+y'^2} - y'\left(y'\sqrt{1+y'^2} + \frac{yy'y''}{\sqrt{1+y'^2}}\right)}{y^2(1+y'^2)}.$$

Subtracting the last two terms and using common denominators, the Euler-Lagrange differential equation becomes

$$-(1+y'^2)^2 - (yy'' - y'^2 - y'^4) = 0.$$

Further algebra yields

$$-1 - 2y'^2 - y'^4 - yy'' + y'^2 + y'^4 = 0,$$

which shortens to

$$yy'' + y'^2 = -1.$$

Recognizing that

$$(yy')' = y'^2 + yy'',$$

integrating results in

$$yy' = -x + c_1.$$

Separating the variables, we obtain

$$ydy = (-x + c_1)dx,$$

and further integration

$$\frac{1}{2}y^2 = -\frac{x^2}{2} + c_1 x + c_2$$

produces the general (implicit) solution as

$$y^2 = -x^2 + 2c_1 x + 2c_2.$$

The fixed boundary condition on the left enables the resolution of one of the constants of integration

$$y^2(0) = 0 = 2c_2 \rightarrow c_2 = 0$$

and the solution now is of the form

$$y^2 = -x^2 + 2c_1 x.$$

To satisfy the algebraic constraint, we reorganize the transversality condition using the simplified notation as

$$f_{y'}\big|_{x_1} - \frac{f\big|_{x_1}}{\dfrac{g_x\big|_{x_1}}{g_y\big|_{y_1}} + y'\big|_{x_1}} = 0.$$

Multiplying and reordering yields a computationally expedient form

$$\left(f - (\frac{g_x}{g_y} + y')f_{y'}\right)\Big|_{x_1} = 0,$$

where the posted substitution at x_1 was moved outside for simplicity of notation. The derivatives of the constraint equation are

$$g_y = 1; g_x = -1$$

and their ratio is also -1. The transversality condition for this problem then becomes

$$\left(f - (-1 + y')f_{y'}\right)\Big|_{x_1} = 0.$$

In detail

$$\left(\frac{\sqrt{1+y'^2}}{y} + \frac{(1-y')y'}{y\sqrt{1+y'^2}}\right)\Big|_{x_1} = \left(1 + y'^2 + y' - y'^2\right)\Big|_{x_1} = 1 + y'(x_1) = 0$$

hence

$$y'(x_1) = -1.$$

Since the solution from the Euler-Lagrange equation was

$$y^2 = -x^2 + 2c_1 x,$$

differentiating produces

$$2yy' = -2x + 2c_1.$$

Substituting x_1 results in

$$2y(x_1)y'(x_1) = -2x_1 + 2c_1.$$

The solution at that point must also satisfy the constraint equation

$$y(x_1) = x_1 - 5;$$

therefore, the equation

$$2(x_1 - 5)(-1) = -2x_1 + 2c_1$$

emerges from which the unknown coefficient is easily solved as

$$c_1 = 5.$$

Hence, the specific implicit solution of the problem is

$$y^2 = -x^2 + 10x,$$

or explicitly

$$y = \sqrt{-x^2 + 10x}.$$

Intersecting this with the constraint curve, the actual values of x_1, y_1 may be obtained if necessary.

_____

2.2   Lagrange's solution

We now further generalize the variational problem and impose both boundary conditions as well as an algebraic condition on the whole domain as follows:

$$I(y) = \int_{x_0}^{x_1} f(x, y, y')dx = \text{extremum},$$

with

$$y(x_0) = y_0, y(x_1) = y_1,$$

while

$$J(y) = \int_{x_0}^{x_1} g(x, y, y')dx = \text{constant}.$$

Following the earlier established pattern, we introduce an alternative solution function, at this time, however, with two auxiliary functions as

$$Y(x) = y(x) + \epsilon_1 \eta_1(x) + \epsilon_2 \eta_2(x).$$

Here the two auxiliary functions are arbitrary and both satisfy the conditions:

$$\eta_1(x_0) = \eta_1(x_1) = \eta_2(x_0) = \eta_2(x_1) = 0.$$

Substituting these into the integrals gives

$$I(Y) = \int_{x_0}^{x_1} f(x, Y, Y')dx,$$

and

$$J(Y) = \int_{x_0}^{x_1} g(x, Y, Y')dx.$$

Lagrange's ingenious solution is to tie the two integrals together with a yet unknown multiplier (called the Lagrange multiplier, λ) as follows:

$$I(\epsilon_1, \epsilon_2) = I(Y) + \lambda J(Y) = \int_{x_0}^{x_1} h(x, Y, Y') dx,$$

where

$$h(x, y, y') = f(x, y, y') + \lambda g(x, y, y').$$

The condition to solve this variational problem is

$$\frac{\partial I}{\partial \epsilon_i} = 0$$

when

$$\epsilon_i = 0; i = 1, 2.$$

The derivatives are of the form

$$\frac{\partial I}{\partial \epsilon_i} = \int_{x_0}^{x_1} \left(\frac{\partial h}{\partial Y} \eta_i + \frac{\partial h}{\partial Y'} \eta_i' \right) dx.$$

The extremum is obtained when

$$\frac{\partial I}{\partial \epsilon_i} \bigg|_{\epsilon_i = 0, i=1,2} = \int_{x_0}^{x_1} \left(\frac{\partial h}{\partial Y} \eta_i + \frac{\partial h}{\partial Y'} \eta_i' \right) dx = 0.$$

Considering the boundary conditions and integrating by parts yield

$$\int_{x_0}^{x_1} \left(\frac{\partial h}{\partial y} - \frac{d}{dx} \frac{\partial h}{\partial y'} \right) \eta_i dx = 0,$$

which, due to the fundamental lemma of calculus of variations, results in the relevant Euler-Lagrange differential equation

$$\frac{\partial h}{\partial y} - \frac{d}{dx} \frac{\partial h}{\partial y'} = 0.$$

This equation contains three undefined coefficients: the two coefficients of integration satisfying the boundary conditions and the Lagrange multiplier, enforcing the constraint.

2.3 Isoperimetric problems

Isoperimetric problems use a given perimeter of a certain object as the constraint of some variational problem. The perimeter may be a curve in the two-dimensional case, as in the example of the next section. It may also be the surface of a certain body in the three-dimensional case.

2.3.1 Maximal area under curve with given length

This problem is conceptually very simple, but useful to illuminate the process just established. It is also a very practical problem with more difficult geometries involved. Here we focus on the simple case of finding the curve of given length between two points in the plane. Without restricting the generality of the discussion, we will position the two points on the x axis in order to simplify the algebraic work.

The given points are $(x_0, 0)$ and $(x_1, 0)$ with $x_0 < x_1$. Note, however, that their distance is not specified, a fact to be addressed later. The area under any curve going from the fixed start point x_0 to the open endpoint x_1 in the upper half-plane is

$$I(y) = \int_{x_0}^{x_1} y\, dx.$$

The constraint of the given length L is presented by the equation

$$J(y) = \int_{x_0}^{x_1} \sqrt{1 + y'^2}\, dx = L.$$

The Lagrange multiplier method brings the function

$$h(x, y, y') = y + \lambda\sqrt{1 + y'^2}.$$

The constrained variational problem is

$$I(y) = \int_{x_0}^{x_1} h(x, y, y')\, dx = \text{extremum},$$

whose Euler-Lagrange equation becomes

$$1 - \lambda \frac{d}{dx} \frac{y'}{\sqrt{1 + y'^2}} = 0.$$

Reordering as

$$\frac{d}{dx} \frac{\lambda y'}{\sqrt{1 + y'^2}} = 1$$

yields

$$\frac{\lambda y'}{\sqrt{1 + y'^2}} = x - c_1,$$

where the sign of the integrating constant was chosen for further convenience. The form hints that, instead of differentiating, a simple integration will suffice.

The algebraic steps of

$$\frac{\lambda^2 y'^2}{1 + y'^2} = (x - c_1)^2,$$

$$\lambda^2 y'^2 = (x - c_1)^2 + y'^2(x - c_1)^2,$$

and

$$y'^2(\lambda^2 - (x - c_1)^2) = (x - c_1)^2$$

produce

$$y' = \frac{x - c_1}{\sqrt{\lambda^2 - (x - c_1)^2}}.$$

Finally we separate the variables as

$$dy = \frac{x - c_1}{\sqrt{\lambda^2 - (x - c_1)^2}} dx,$$

and integrate again to produce

$$y(x) = \sqrt{\lambda^2 - (x - c_1)^2} + c_2.$$

It is easy to reorder this into

$$(x - c_1)^2 + (y - c_2)^2 = \lambda^2,$$

which is the equation of a circle. Since the two given points are on the x axis, the center of the circle must lie on the perpendicular bisector of the chord, which implies that

$$c_1 = \frac{x_0 + x_1}{2}.$$

To solve for the value of the Lagrange multiplier (that is the radius) and the other constant, we consider that the circular arc between the two points is the given length:

$$L = \lambda\theta,$$

where θ is the angle of the arc. The angle is related to the remaining constant as

$$2\pi - \theta = 2\arctan\left(\frac{x_1 - x_0}{2c_2}\right).$$

The two equations may be simultaneously satisfied with

$$\theta = \pi,$$

resulting in the shape being a semi-circle. This yields the solutions of

$$c_2 = 0$$

and
$$\lambda = \frac{L}{\pi}.$$

The final solution function in implicit form is

$$\left(x - \frac{x_0 + x_1}{2}\right)^2 + y^2 = \left(\frac{L}{\pi}\right)^2,$$

or explicitly

$$y(x) = \sqrt{\left(\frac{L}{\pi}\right)^2 - \left(x - \frac{x_0 + x_1}{2}\right)^2}.$$

The consequence of this result is that the distance of the open boundary point x_1, to the fixed one, must be the diameter of the circle. Since the radius is L/π, the location of the open boundary point becomes

$$x_1 = x_0 + 2\frac{L}{\pi}.$$

For example
$$x_0 = 0,$$

and
$$L = \frac{\pi}{2}$$

would result in
$$c_1 = 0.5$$

and
$$x_1 = 1.0$$

as shown Figure 2.1.

It is simple to verify that the solution produces the extremum of the original variational problem. Figure 2.1 visibly demonstrates the phenomenon with three curves of equal length $(\pi/2)$ over the same interval. Neither of the solid curves denoted by $g(x)$, the triangle or the rectangle, cover as much area as the semi-circle $y(x)$ marked by the dashed lines.

On the other hand, since this was really a constrained boundary problem, the method developed in Section 2.1 may also be applied. The condition

$$y(x_1) = 0$$

represents an algebraic constraint in the form of

$$g(x, y) = y = 0.$$

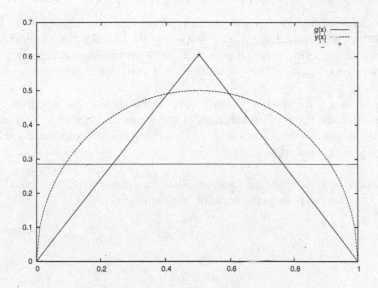

FIGURE 2.1 Maximum area under curves

The first part of the solution, the Euler-Lagrange differential equation, is the same as above. The transversality condition defined for such scenario at the end of Section 2.1 is applied for this case as

$$\frac{\lambda y'}{\sqrt{1+y'^2}} - \frac{y + \lambda\sqrt{1+y'^2}}{y'}\bigg|_{x=x_1} = 0.$$

Since $y(x_1) = 0$ this simplifies to

$$\lambda\frac{-1}{y'\sqrt{1+y'^2}} = 0.$$

Since λ cannot be zero,

$$\frac{1}{y'\sqrt{1+y'^2}} = 0$$

that results in

$$y'(x_1) = \infty.$$

Geometrically this means that the curve has a vertical tangent at the end point, implying that it must be a half circle. Hence, the solutions are in agreement.

2.3.2 Optimal shape of curve of given length under gravity

Another constrained variational problem, whose final result is often used in engineering, is the rope hanging under its weight. The practical importance of the problem regarding power lines and suspended cables is well-known. Here we derive the solution of this problem from a variational origin.

A body in a force field is in static equilibrium when its potential energy has a stationary value. Furthermore, if the stationary value is a minimum, then the body is in stable equilibrium. This is known as principle of minimum potential energy and also originated by Euler.

Assume a body of a homogeneous cable with a given weight per unit length of $\rho =$ constant, and suspension point locations of

$$P_0 = (x_0, y_0),$$

and

$$P_1 = (x_1, y_1).$$

These constitute the boundary conditions. A constraint is also given on the length of the curve: L. The potential energy of the cable is

$$E_p = \int_{P_0}^{P_1} \rho y ds,$$

where y is the distance of the infinitesimal arc segment from the horizontal base line and ρds is its weight. Using the arc length formula, we obtain

$$E_p = \rho \int_{x_0}^{x_1} y\sqrt{1 + y'^2}dx.$$

The principle of minimal potential energy dictates that the equilibrium position of the cable is the solution of the variational problem of

$$I(y) = \rho \int_{x_0}^{x_1} y\sqrt{1 + y'^2}dx = \text{extremum},$$

under boundary conditions

$$y(x_0) = y_0; y(x_1) = y_1$$

and constraint of

$$\int_{x_0}^{x_1} \sqrt{1 + y'^2}dx = L.$$

Introducing the Lagrange multiplier and the constrained function

$$h(y) = \rho y\sqrt{1 + y'^2} + \lambda\sqrt{1 + y'^2},$$

or

$$h(y) = (\rho y + \lambda)\sqrt{1 + y'^2},$$

the Euler-Lagrange differential equation may be obtained by Beltrami's formula since the independent variable does not exist explicitly in the functional. Hence

$$(\rho y + \lambda)\sqrt{1 + y'^2} - \frac{(\rho y + \lambda)y'^2}{\sqrt{1 + y'^2}} = c_1,$$

where c_1 is the constant of integration. This simplifies to

$$\frac{(\rho y + \lambda)}{\sqrt{1 + y'^2}} = c_1$$

and

$$(\rho y + \lambda)^2 = c_1^2(1 + y'^2).$$

Expressing the derivative

$$y' = \frac{\sqrt{(\rho y + \lambda)^2 - c_1^2}}{c_1},$$

separating brings

$$\frac{c_1\,dy}{\sqrt{(\rho y + \lambda)^2 - c_1^2}} = dx.$$

Substituting

$$z = \rho y + \lambda,$$

and posting the integrals result in

$$\frac{c_1}{\rho} \int \frac{dz}{\sqrt{z^2 - c_1^2}} = \int dx.$$

Executing the integrals, we obtain

$$\frac{c_1}{\rho} \cosh^{-1}\left(\frac{z}{c_1}\right) + c_2 = x$$

with c_2 being another constant of integration. Then

$$\cosh^{-1}\left(\frac{z}{c_1}\right) = \frac{\rho}{c_1}(x - c_2),$$

taking the inverse of the hyperbolic function and back-substituting brings

$$\frac{z}{c_1} = \frac{\rho y + \lambda}{c_1} = \cosh\left(\frac{\rho(x - c_2)}{c_1}\right).$$

From this the solution of the so-called catenary curve emerges

$$y = -\frac{\lambda}{\rho} + \frac{c_1}{\rho} \cosh\left(\frac{\rho(x - c_2)}{c_1}\right).$$

The constants of integration may be determined by the boundary conditions albeit the calculation, due to the presence of the hyperbolic function, is rather tedious. Let us consider the specific case of the suspension points being at the same height and symmetric with respect to the origin. This is a typical engineering scenario for the span of suspension cables. This results in the following boundary conditions:

$$P_0 = (x_0, y_0) = (-s, h)$$

and

$$P_1 = (x_1, y_1) = (s, h).$$

Without the loss of the generality, we can consider unit weight ($\rho = 1$) and by substituting above boundary conditions, we obtain

$$h + \lambda = c_1 \cosh\left(\frac{-s + c_2}{c_1}\right) = c_1 \cosh\left(\frac{s + c_2}{c_1}\right).$$

This implies that

$$c_2 = 0.$$

The value of the second coefficient is solved by adhering to the length constraint. Integrating the constraint equation yields

$$L = 2c_1 \sinh\left(\frac{s}{c_1}\right)$$

whose only unknown is the integration constant c_1. This problem is not solvable by analytic means; however, it can be solved by an iterative procedure numerically by considering the unknown coefficient as a variable:

$$c_1 = x,$$

and intersecting the curve

$$y = x \sinh\left(\frac{s}{x}\right)$$

and the horizontal line

$$y = \frac{L}{2}.$$

The minimal cable length must exceed the width of the span; hence, we expect the cable to have some slack. Then, for example, using

$$L = 3s$$

will result in an approximate solution of

$$c_1 = 0.6175.$$

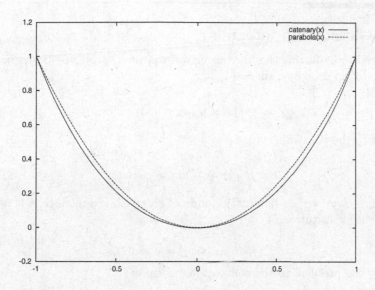

FIGURE 2.2 The catenary curve

Clearly, depending on the length of the cable between similarly posted suspension locations, different catenary curves may be obtained.

The Lagrange multiplier may finally be resolved by the expression

$$\lambda = c_1 \cosh\left(\frac{s}{c_1}\right) - h.$$

Assuming a cable suspended with a unit half-span ($s = 1$) and from unit height ($h = 1$) and length of three times the half-span ($L = 3$), the value of the Lagrange multiplier becomes

$$\lambda = 0.6175 \ \cosh\left(\frac{1}{0.6175}\right) - 1 = 0.6204.$$

The final catenary solution curve, shown with a solid line in Figure 2.2, is represented by

$$y = 0.6175 \ \cosh\left(\frac{1}{0.6175}\right) - 0.6204.$$

For comparison purposes, the figure also shows a parabola with dashed lines, representing an approximation of the catenary and obeying the same boundary conditions.

2.4   Closed-loop integrals

As a final topic in this chapter, we view variational problems posed in terms of closed-loop integrals, such as

$$I = \oint f(x, y, y') dx = \text{extremum},$$

subject to the constraint of

$$J = \oint g(x, y, y') dx = \text{constant}.$$

Note that there are no boundary points of the path given, since it is a closed loop. The substitution of

$$x = a \, \cos(t), \; y = a \, \sin(t),$$

changes the problem to the conventional form of

$$I = \int_{t_0}^{t_1} F(x, y, \dot{x}, \dot{y}) dt = \text{extremum},$$

subject to

$$J = \int_{t_0}^{t_1} G(x, y, \dot{x}, \dot{y}) dt = \text{constant}.$$

The arbitrary t_0 and the specific $t_1 = t_0 + 2\pi$ boundary points clearly cover a complete loop. The change resulted in a variational form of two parametric variables, a topic of the next chapter.

To illustrate the solution with closed-loop integrals, we attempt to find the curve with given arc length enclosing maximum area. The problem may be posted as

$$\int \int_D dA = \text{extremum}$$

under the constraint presented by the closed-loop integral of

$$\oint_C ds = L.$$

Here D is the domain enclosed by the curve C. Using Green's identity of

$$\int \int_D dA = \frac{1}{2} \oint_C x dy - y dx = \frac{1}{2} \oint_C (xy' - y) dx,$$

and substituting the arc length, we restate the problem as

$$\frac{1}{2} \oint (xy' - y) dx = \text{extremum},$$

with the constraint

$$\oint \sqrt{1 + y'^2}\, dx = L.$$

Using the Lagrange multiplier approach introduced in this chapter, the functional becomes

$$I = \oint_C \left(\frac{1}{2}(xy' - y) + \lambda\sqrt{1 + y'^2} \right) dx = \text{extremum}.$$

The components of the Euler-Lagrange differential equation for this case are

$$\frac{\partial h}{\partial y} = -\frac{1}{2},$$

and

$$\frac{d}{dx}\frac{\partial h}{\partial y'} = \frac{d}{dx}\left(\frac{1}{2}x + \frac{\lambda y'}{\sqrt{1 + y'^2}} \right).$$

Differentiating only the first term in the latter and combining with the other component becomes

$$\frac{d}{dx}\left(\frac{\lambda y'}{\sqrt{1 + y'^2}} \right) = -1.$$

Integrating both sides brings

$$\frac{\lambda y'}{\sqrt{1 + y'^2}} = -x + c_1.$$

Expressing the derivative function results in

$$y' = \frac{x - c_1}{\sqrt{\lambda^2 - (x - c_1)^2}}.$$

Finally integrating produces the explicit solution of

$$y = \sqrt{\lambda^2 - (x - c_1)^2} + c_2.$$

In this case, the implicit solution is more intuitive as

$$(x - c_1)^2 + (y - c_2)^2 = \lambda^2,$$

which is clearly a closed circle. Specifying boundary conditions would enable us to resolve the location of the center points captured in c_1, c_2 and the given value L of the constraint would specify the radius hidden in the Lagrange multiplier λ.

Note that this problem may also be stated in a dual form as to find a curve with minimum length enclosing a given area in the plane. The functional in this case becomes

$$I = \oint_C \left(\sqrt{1 + y'^2} + \lambda \left(\frac{1}{2}(xy' - y) \right) \right) dx = \text{extremum}.$$

Dividing by the constant λ produces the identical Euler-Lagrange differential equation with a different (undefined) Lagrange multiplier. Hence, not surprisingly, the solution is again the circle with two integration constants defining the center point and a Lagrange multiplier related radius.

2.5 Exercises

Find the solutions for the problems by using the Euler-Lagrange differential equation and the boundary conditions and cosntraints, if given.

1.
$I = \int_0^{x_1} \frac{\sqrt{1+y'^2}}{y} dx = $ extremum.
Boundary conditions: $y(0) = 0, (x_1, y_1) \in (x-9)^2 + y^2 = 9$.

2.
$I = \int_0^1 (y'^2 + x^2) dx = $ extremum.
Constraint: $\int_0^1 y^2 dx = 2$.

3.
$I = \int_0^\pi y'^2 dx = $ extremum.
Constraint: $\int_0^\pi y^2 dx = 1$.

4.
$I = \int_0^\pi (y'^2 - y^2) dx = $ extremum.
Constraint: $\int_0^\pi y dx = 1$.

5.
$I = \oint \sqrt{1 + y'^2} dx = $ extremum.
Constraint: $\frac{1}{2} \oint (xy' - y) dx = A$.

6.
$I = 2\pi \int_{x_1}^{x_2} y \sqrt{1 + y'^2} dx = $ extremum.
Constraint: $\int_{x_1}^{x_2} \sqrt{1 + y'^2} dx = L$.

7.
$I = \int_0^{\pi/2} (y'^2 - y^2) dx = $ extremum.
Boundary conditions: $y(0) = 0, y(\pi/2) = 1$.
Constraint: $\int_0^{\pi/2} 2y dx = 6 - \pi$.

8.

$I = \int_0^1 (y'^2 + xy)dx = $ extremum.

Boundary conditions: $y(0) = -\frac{1}{12}, y(1) = 2$.

Constraint: $\int_0^1 12y\,dx = 6$.

9.

$I = \int_0^{x_1} (y'^2 + x^2)dx = $ extremum.

Boundary conditions: $y(0) = 0, (x_1, y_1) \in y = 4 - x^2$.

10.

$I = \int_0^1 (y'^2)dx = $ extremum.

Boundary conditions: $y(0) = 0, y(1) = 2$.

Constraint: $\int_0^1 y\,dx = 4$.

3

Multivariate functionals

The subject of this chapter is further expansion of the classes of functions used in variational problems. Specifically, we discuss functionals with multiple functions, functions with multiple variables will be discussed in a variational context. Parametric functions and minimal surface applications will also be presented.

3.1 Functionals with several functions

The variational problem of multiple dependent variables is posed as

$$I(y_1, y_2, \ldots, y_n) = \int_{x_0}^{x_1} f(x, y_1, y_2, \ldots, y_n, y_1', y_2', \ldots, y_n') dx$$

with a pair of boundary conditions given for all functions:

$$y_i(x_0) = y_{i,0}$$

and

$$y_i(x_1) = y_{i,1}$$

for each $i = 1, 2, \ldots, n$. The alternative solutions are:

$$Y_i(x) = y_i(x) + \epsilon_i \eta_i(x); i = 1, 2, \ldots, n$$

with all the arbitrary auxiliary functions obeying the conditions:

$$\eta_i(x_0) = \eta_i(x_1) = 0.$$

The variational problem becomes

$$I(\epsilon_1, \ldots, \epsilon_n) = \int_{x_0}^{x_1} f(x, \ldots, y_i + \epsilon_i \eta_i, \ldots, y_i' + \epsilon_i \eta_i', \ldots) dx,$$

whose derivative with respect to the auxiliary variables is

$$\frac{\partial I}{\partial \epsilon_i} = \int_{x_0}^{x_1} \frac{\partial f}{\partial \epsilon_i} dx = 0.$$

Applying the chain rule we get

$$\frac{\partial f}{\partial \epsilon_i} = \frac{\partial f}{\partial Y_i}\frac{\partial Y_i}{\partial \epsilon_i} + \frac{\partial f}{\partial Y_i'}\frac{\partial Y_i'}{\partial \epsilon_i} = \frac{\partial f}{\partial Y_i}\eta_i + \frac{\partial f}{\partial Y_i'}\eta_i'.$$

Substituting into the variational equation yields, for $i = 1, 2, \ldots, n$:

$$\frac{\partial I}{\partial \epsilon_i} = \int_{x_0}^{x_1}\left(\frac{\partial f}{\partial Y_i}\eta_i + \frac{\partial f}{\partial Y_i'}\eta_i'\right) dx.$$

Integrating by parts and exploiting the alternative function form results in

$$\frac{\partial I}{\partial \epsilon_i} = \int_{x_0}^{x_1}\eta_i\left(\frac{\partial f}{\partial y_i} - \frac{d}{dx}\frac{\partial f}{\partial y_i'}\right) dx.$$

To reach the extremum, based on the fundamental lemma, we need the solution of a set of n Euler-Lagrange equations of the form

$$\frac{\partial f}{\partial y_i} - \frac{d}{dx}\frac{\partial f}{\partial y_i'} = 0; i = 1, \ldots, n.$$

Note that in this case the separate functions were dependent on the same parameter. A later section discusses the case of one dependent function being turned into this class of problems by introducing a parametric representation.

3.1.1   Euler–Lagrange system of equations

To illustrate this scenario, we consider an example where two functions are dependent on a common independent variable,

$$\int_{x_0}^{x_1}\left(2y_1y_2 + \left(\frac{dy_1}{dx}\right)^2 + \left(\frac{dy_2}{dx}\right)^2\right) dx = \text{extremum},$$

with the following boundary conditions at

$$x_0 = 0, x_1 = \frac{\pi}{2}.$$

In order to solve the problem, we need distinct boundary conditions for the two functions:

$$y_1(x_0) = 0, y_2(x_0) = 0,$$

and

$$y_1(x_1) = -1, y_2(x_1) = 1.$$

According to the formula developed above, we have a system of two Euler-Lagrange equations

$$\frac{\partial f}{\partial y_1} - \frac{d}{dx}\frac{\partial f}{\partial y_1'} = 0$$

and

$$\frac{\partial f}{\partial y_2} - \frac{d}{dx}\frac{\partial f}{\partial y_2'} = 0.$$

Specifically for our functional

$$2y_2 - \frac{d}{dx}(2y_1') = 0$$

and

$$2y_1 - \frac{d}{dx}(2y_2') = 0.$$

Differentiation and cancellation produce the coupled system of differential equations

$$\frac{d^2 y_1}{dx^2} = y_2$$

and

$$\frac{d^2 y_2}{dx^2} = y_1.$$

Substituting the first equation into the second as

$$\frac{d^2}{dx^2}\frac{d^2 y_1}{dx^2} = y_1$$

produces the fourth-order homogeneous equation

$$\frac{d^4 y_1}{dx^4} - y_1 = 0.$$

The characteristic equation of

$$\lambda^4 - 1 = 0$$

produces the solutions of

$$1, -1, i, -i.$$

The corresponding solutions using Euler's formula to convert the complex terms into reals are

$$y_1(x) = ae^x + be^{-x} + c\cos(x) + d\sin(x),$$

and since

$$y_2 = \frac{d^2 y_1}{dx^2},$$

the other solution becomes

$$y_2(x) = ae^x + be^{-x} - c\cos(x) - d\sin(x),$$

Applying the boundary conditions at the lower end, we obtain

$$y_1(0) = ae^0 + be^0 + c\cos(0) + d\sin(0) = a + b + c = 0$$

and
$$y_2(0) = ae^0 + be^0 - c\cos(0) - d\sin(0) = a + b - c = 0.$$

Similarly at the upper end

$$y_1\left(\frac{\pi}{2}\right) = ae^{\frac{\pi}{2}} + be^{\frac{-\pi}{2}} + c\cos\left(\frac{\pi}{2}\right) + d\sin\left(\frac{\pi}{2}\right) = ae^{\frac{\pi}{2}} + be^{\frac{-\pi}{2}} + d = -1$$

and

$$y_2\left(\frac{\pi}{2}\right) = ae^{\frac{\pi}{2}} - be^{\frac{-\pi}{2}} - c\cos\left(\frac{\pi}{2}\right) - d\sin\left(\frac{\pi}{2}\right) = ae^{\frac{\pi}{2}} + be^{\frac{-\pi}{2}} - d = 1.$$

Without details of the elementary algebra, we conclude that

$$a = 0, b = 0, c = 0, d = -1.$$

Hence the solution to the problem is

$$y_1 = -\sin(x),$$

and from their relationship of

$$y_2 = \frac{d^2 y_1}{dx^2}$$

the other solution becomes

$$y_2 = \sin(x).$$

3.2   Variational problems in parametric form

Most of the discussion insofar was focused on functions in explicit form. The concepts also apply to problems posed in parametric form. The explicit form variational problem of

$$I(y) = \int_{x_0}^{x_1} f(x, y, y')dx$$

may be reformulated with the substitutions

$$x = u(t), \; y = v(t).$$

The parametric variational problem becomes of the form

$$I(x, y) = \int_{t_0}^{t_1} f(x, y, \frac{\dot{y}}{\dot{x}})\dot{x}\,dt,$$

or

$$I(x, y) = \int_{t_0}^{t_1} F(t, x, y, \dot{x}, \dot{y})dt.$$

The Euler-Lagrange differential equation system for this case becomes

$$\frac{\partial F}{\partial x} - \frac{d}{dt}\frac{\partial F}{\partial \dot{x}} = 0,$$

and

$$\frac{\partial F}{\partial y} - \frac{d}{dt}\frac{\partial F}{\partial \dot{y}} = 0.$$

It is proven in [6] that an explicit variational problem is invariant under parameterization. In other words, regardless of the algebraic form of the parameterization, the same explicit solution will be obtained.

Parametrically given problems may be considered as functionals with several functions. As an example, we consider the following twice differentiable functions

$$x = x(t), y = y(t), z = z(t).$$

The variational problem in this case is presented as

$$I(x,y,z) = \int_{t_0}^{t_1} f(t,x,y,z,\dot{x},\dot{y},\dot{z})dt.$$

Here the independent variable t is the parameter, and there are three dependent variables : x, y, z. Applying the steps just explained for this specific case results in the system of Euler-Lagrange equations

$$\frac{\partial f}{\partial x} - \frac{d}{dt}\frac{\partial f}{\partial \dot{x}} = 0,$$

$$\frac{\partial f}{\partial y} - \frac{d}{dt}\frac{\partial f}{\partial \dot{y}} = 0,$$

and

$$\frac{\partial f}{\partial z} - \frac{d}{dt}\frac{\partial f}{\partial \dot{z}} = 0.$$

To illustrate this scenario, we consider the problem

$$I = \int_0^1 ((\dot{x})^2 + (\dot{y})^2 + 2x)dt = \text{extremum}$$

consisting of two functions $x(t), y(t)$ dependent upon the same parameter t and the boundary conditions

$$x(0) = 1, y(0) = 1; x(1) = \frac{3}{2}, y(1) = 2.$$

The components of the Euler-Lagrange differential equations are

$$\frac{\partial f}{\partial x} = 2,$$

$$\frac{\partial f}{\partial \dot{x}} = 2\dot{x},$$

and

$$\frac{\partial f}{\partial y} = 0,$$

$$\frac{\partial f}{\partial \dot{y}} = 2\dot{y}.$$

The system of Euler-Lagrange differential equations becomes

$$2 - \frac{d}{dt}(2\dot{x}) = 0,$$

$$-\frac{d}{dt}(2\dot{y}) = 0.$$

The equations are decoupled and may be solved separately. The first one after shortening and reordering is

$$\ddot{x} = 1.$$

Integrating both sides yields

$$\dot{x} = t + c_1;$$

and repeating produces the first dependent variable solution

$$x(t) = \frac{t^2}{2} + c_1 t + c_2.$$

Similar activity on the second equation proceeds as

$$\ddot{y} = 0,$$

$$\dot{y} = d_1,$$

and

$$y(t) = d_1 t + d_2.$$

Applying the boundary conditions

$$x(0) = 1 \rightarrow c_2 = 1,$$

and

$$x(1) = \frac{3}{2} = \frac{1}{2} + c_1 + 1 \rightarrow c_1 = 0.$$

Similarly

$$y(0) = 1 \rightarrow d_2 = 1,$$

and

$$y(1) = 2 = d_1 \cdot 1 + 1 \rightarrow d_1 = 1.$$

Finally, the solutions are

$$x(t) = \frac{t^2}{2} + 1,$$

and

$$y(t) = t + 1.$$

Note that the parametric Euler-Lagrange systems are not necessarily decoupled as was shown in Section 3.1.1, contrary to this example.

3.2.1 Maximal area by closed parametric curve

The problem is to find the not self-intersecting closed parametric curve with a given length enclosing maximum area in two dimensions. This problem was also addressed in the last chapter in a single function form.

We use Green's identity to establish the area in parametric form as

$$A = \frac{1}{2} \int x dy - y dx = \frac{1}{2} \int (x \frac{dy}{dt} - y \frac{dx}{dt}) dt = \frac{1}{2} \int (x\dot{y} - y\dot{x}) dt.$$

The length of the parametric curve is

$$L = \int_{t_0}^{t_1} \sqrt{(\dot{x})^2 + (\dot{y})^2} dt.$$

The closed nature of the curve is expressed with the boundary conditions

$$x(t_0) = x(t_1); y(t_0) = y(t_1).$$

The conditional variational statement expressing the problem at hand is

$$I = \int_{t_0}^{t_1} \frac{1}{2}(x\dot{y} - y\dot{x}) + \lambda\sqrt{(\dot{x})^2 + (\dot{y})^2} dt = \text{extremum}.$$

The Euler-Lagrange differential equations become

$$\frac{1}{2}\dot{y} - \frac{d}{dt}\left(-\frac{1}{2}y + \frac{1}{2}\frac{2\lambda\dot{x}}{\sqrt{(\dot{x})^2 + (\dot{y})^2}} \right) = 0$$

and

$$-\frac{1}{2}\dot{x} - \frac{d}{dt}\left(\frac{1}{2}x + \frac{1}{2}\frac{2\lambda\dot{y}}{\sqrt{(\dot{x})^2 + (\dot{y})^2}} \right) = 0.$$

Executing the differentiation only on the first terms in the parenthesis results in

$$\dot{y} - \frac{d}{dt}\left(\frac{\lambda\dot{x}}{\sqrt{(\dot{x})^2 + (\dot{y})^2}} \right) = 0$$

and

$$-\dot{x} - \frac{d}{dt}\left(\frac{\lambda\dot{y}}{\sqrt{(\dot{x})^2 + (\dot{y})^2}} \right) = 0.$$

Integrating both equations, we obtain

$$y - \frac{\lambda\dot{x}}{\sqrt{(\dot{x})^2 + (\dot{y})^2}} = c_2$$

and

$$x + \frac{\lambda\dot{y}}{\sqrt{(\dot{x})^2 + (\dot{y})^2}} = c_1.$$

For convenience, we return to the explicit form as

$$\frac{\dot{y}}{\dot{x}} = \frac{dy}{dx} = y'$$

and substituting into the second pair of equations we get

$$x - c_1 = \frac{-\lambda y'}{\sqrt{1 + y'^2}}.$$

Squaring both sides and multiplying results in

$$(x - c_1)^2 (1 + y'^2) = \lambda^2 y'^2$$

from which the derivative is obtained as

$$y' = \frac{x - c_1}{\sqrt{\lambda^2 - (x - c_1)^2}}.$$

Integrating both sides produces

$$y = \sqrt{\lambda^2 - (x - c_1)^2} + c_2.$$

This is clearly a circle as

$$(y - c_2)^2 + (x - c_1)^2 = \lambda^2.$$

From the given length of L and the fact that it is a closed circle, the resolution of the Lagrange multiplier brings

$$\lambda = \frac{L}{2\pi}.$$

The remaining integrating constants, representing the coordinates of the circle, may be resolved from the equations

$$(y_0 - c_2)^2 + (x_0 - c_1)^2 = \left(\frac{L}{2\pi}\right)^2$$

and

$$(y_1 - c_2)^2 + (x_1 - c_1)^2 = \left(\frac{L}{2\pi}\right)^2.$$

In order to do so, two distinct $(t_0 \neq t_1)$ boundary conditions are required

$$x(t_0) = x_0, y(t_0) = y_0,$$

and

$$x(t_1) = x_1, y(t_1) = y_1.$$

The most practical applications in this class occur in three-dimensional geometry problems to be explored in Chapters 8 and 9.

3.3   Functionals with two independent variables

All our discussions so far were confined to a single integral of the functional. The next step of generalization is to allow a functional with multiple independent variables. The simplest case is that of two independent variables, and this will be the vehicle to introduce the process. The problem is of the form

$$\dot{I}(z) = \int_{y_0}^{y_1} \int_{x_0}^{x_1} f(x, y, z, z_x, z_y) dx dy = \text{extremum}.$$

Here the derivatives are

$$z_x = \frac{\partial z}{\partial x}$$

and

$$z_y = \frac{\partial z}{\partial y}.$$

The alternative solution is also a function of two variables

$$Z(x, y) = z(x, y) + \epsilon \eta(x, y).$$

The now familiar process emerges as

$$I(\epsilon) = \int_{y_0}^{y_1} \int_{x_0}^{x_1} f(x, y, Z, Z_x, Z_y) dx dy = \text{extremum}.$$

The extremum is obtained via the derivative

$$\frac{\partial I}{\partial \epsilon} = \int_{y_0}^{y_1} \int_{x_0}^{x_1} \frac{\partial f}{\partial \epsilon} dx dy.$$

Differentiating and substituting yield

$$\frac{\partial I}{\partial \epsilon} = \int_{y_0}^{y_1} \int_{x_0}^{x_1} \left(\frac{\partial f}{\partial Z} \eta + \frac{\partial f}{\partial Z_x} \eta_x + \frac{\partial f}{\partial Z_y} \eta_y \right) dx dy.$$

The extremum is reached when $\epsilon = 0$:

$$\frac{\partial I}{\partial \epsilon} \Big|_{\epsilon=0} = \int_{y_0}^{y_1} \int_{x_0}^{x_1} \left(\frac{\partial f}{\partial z} \eta + \frac{\partial f}{\partial z_x} \eta_x + \frac{\partial f}{\partial z_y} \eta_y \right) dx dy = 0.$$

Applying Green's identity for the second and third terms produces

$$\int_{y_0}^{y_1} \int_{x_0}^{x_1} \left(\frac{\partial f}{\partial z} - \frac{\partial}{\partial x} \frac{\partial f}{\partial z_x} - \frac{\partial}{\partial y} \frac{\partial f}{\partial z_y} \right) \eta dx dy + \int_{\partial D} \left(\frac{\partial f}{\partial z_x} \frac{dy}{ds} - \frac{\partial f}{\partial z_y} \frac{dx}{ds} \right) \eta ds = 0.$$

Here ∂D is the boundary of the domain of the problem, and the second integral vanishes by the definition of the auxiliary function. Due to the fundamental lemma of calculus of variations, the Euler-Lagrange differential equation becomes

$$\frac{\partial f}{\partial z} - \frac{\partial}{\partial x}\frac{\partial f}{\partial z_x} - \frac{\partial}{\partial y}\frac{\partial f}{\partial z_y} = 0.$$

3.4   Minimal surfaces

Minimal surfaces have intriguing natural occurrences. For example, soap films stretched over various types of wire loops intrinsically attain such shapes, no matter how difficult the boundary curve is. Various biological cell interactions also manifest similar phenomena.

From a differential geometry point-of-view, a minimal surface is a surface for which the mean curvature of the form

$$\kappa_m = \frac{\kappa_1 + \kappa_2}{2}$$

vanishes, where κ_1 and κ_2 are the principal curvatures. A subset of minimal surfaces are the surfaces of minimal area, and surfaces of minimal area passing through a closed space curve are minimal surfaces. Finding minimal surfaces is called the problem of Plateau.

We seek the surface of minimal area with equation

$$z = f(x, y), (x, y) \in D,$$

with a closed-loop boundary curve

$$g(x, y, z) = 0; (x, y) \in \partial D.$$

The boundary condition represents a three-dimensional curve defined over the perimeter of the domain. The curve may be piecewise differentiable, but continuous and forms a closed loop, a Jordan curve.

The corresponding variational problem is

$$I(z) = \int\int_D \sqrt{1 + \frac{\partial z}{\partial x}^2 + \frac{\partial z}{\partial y}^2}\, dxdy = \text{extremum},$$

subject to the constraint of the boundary condition above. The Euler-Lagrange equation for this case is of the form

$$-\frac{\partial}{\partial x}\frac{\frac{\partial z}{\partial x}}{\sqrt{1 + \left(\frac{\partial z}{\partial x}\right)^2 + \left(\frac{\partial z}{\partial y}\right)^2}} - \frac{\partial}{\partial y}\frac{\frac{\partial z}{\partial y}}{\sqrt{1 + \left(\frac{\partial z}{\partial x}\right)^2 + \left(\frac{\partial z}{\partial y}\right)^2}} = 0.$$

After considerable algebraic work, this equation becomes

$$\left(1+\left(\frac{\partial z}{\partial y}\right)^2\right)\frac{\partial^2 z}{\partial x^2} - 2\frac{\partial z}{\partial x}\frac{\partial z}{\partial y}\frac{\partial^2 z}{\partial x \partial y} + \left(1+\left(\frac{\partial z}{\partial x}\right)^2\right)\frac{\partial^2 z}{\partial y^2} = 0.$$

This is the differential equation of minimal surfaces, originally obtained by Lagrange himself. The equation is mainly of verification value as this is one of the most relevant examples for the need of a numerical solution. Most of the problems of finding minimal surfaces are solved by Ritz type methods, the subject of Chapter 7.

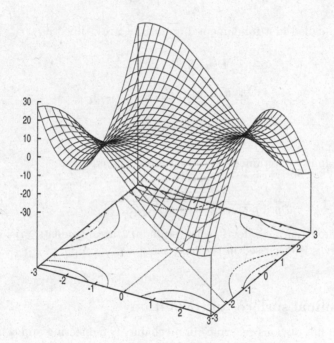

FIGURE 3.1 Saddle surface

The simplest solutions for such problems are the so-called saddle surfaces, such as, for example, shown in Figure 3.1, whose equation is

$$z = x^3 - 2xy^2.$$

It is easy to verify that this satisfies the equation. The figure also shows the level curves of the surface projected to the x-y plane. The straight lines on the plane correspond to geodesic paths, a subject of detailed discussion in Chapter 8. It is apparent that the $x = 0$ planar cross-section of the surface is the $z = 0$ line in the x-y plane, as indicated by the algebra. The intersection with the $y = 0$ plane produces the $z = x^3$ curve, again in full adherence to the equation.

When a minimal surface is sought in a parametric form

$$\underline{r} = x(u, v)\underline{i} + y(u, v)\underline{j} + z(u, v)\underline{k},$$

the variational problem becomes

$$I(\underline{r}) = \int \int_D \sqrt{EF - G^2} dA,$$

where the so-called first fundamental quantities are defined as

$$E(u, v) = |\underline{r}'_u|^2,$$

$$F(u, v) = \underline{r}'_u \cdot \underline{r}'_v,$$

and

$$G(u, v) = |\underline{r}'_v|^2.$$

The solution may be obtained from the differential equation

$$\frac{\partial}{\partial u} \frac{F\underline{r}'_u - G\underline{r}'_v}{\sqrt{EF - G^2}} + \frac{\partial}{\partial v} \frac{E\underline{r}'_v - G\underline{r}'_u}{\sqrt{EF - G^2}} = 0.$$

Finding minimal surfaces for special boundary arrangements arising from revolving curves is discussed in the next section.

3.4.1   Minimal surfaces of revolution

This problem has obvious relevance in mechanical engineering and computer-aided manufacturing (CAM). Let us now consider two points

$$P_0 = (x_0, y_0), P_1 = (x_1, y_1),$$

and find the function $y(x)$ going through the points that generates an object of revolution $z = f(x, y)$ when rotated around the x axis with minimal surface area. The surface of that object of revolution is

$$S = 2\pi \int_{x_0}^{x_1} y\sqrt{1 + y'^2} dx.$$

The corresponding variational problem is

$$I(y) = 2\pi \int_{x_0}^{x_1} y\sqrt{1 + y'^2} dx = \text{extremum},$$

with the boundary conditions of

$$y(x_0) = y_0, y(x_1) = y_1.$$

The Beltrami formula of Equation (1.1) produces

$$y\sqrt{1 + y'^2} - \frac{yy'^2}{\sqrt{1 + y'^2}} = c_1.$$

Reordering and another integration yield

$$x = c_1 \int \frac{1}{\sqrt{y^2 - c_1^2}} dy.$$

Hyperbolic substitution enables the integration as

$$x = c_1 \cosh^{-1}\left(\frac{y}{c_1}\right) + c_2.$$

Finally the solution curve generating the minimal surface of revolution between the two points is

$$y = c_1 \cosh\left(\frac{x - c_2}{c_1}\right),$$

where the integration constants are resolved with the boundary conditions as

$$y_0 = c_1 \cosh\left(\frac{x_0 - c_2}{c_1}\right),$$

and

$$y_1 = c_1 \cosh\left(\frac{x_1 - c_2}{c_1}\right).$$

An example of such a surface of revolution, the catenoid, is shown in Figure 3.2 where the meridian curves are catenary curves.

3.5 Functionals with three independent variables

The generalization to functionals with multiple independent variables is rather straightforward from the last section. The case of three independent variables, however, has such enormous engineering importance that it is worthy of a special section. The problem is of the form

$$I\left(u(x, y, z)\right) = \int\int_D f(x, y, z, u, u_x, u_y, u_z) dx dy dz = \text{extremum}.$$

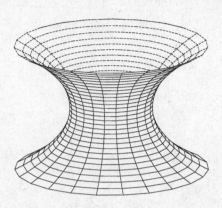

FIGURE 3.2 Catenoid surface

The solution function $u(x, y, z)$ may be some engineering quantity describing a physical phenomenon acting on a three-dimensional body. Here the domain is generalized as well to

$$x_0 \leq x \leq x_1, \ y_0 \leq y \leq y_1, \ z_0 \leq z \leq z_1.$$

The alternative solution is also a function of three variables

$$U(x, y, z) = u(x, y, z) + \epsilon \eta(x, y, z).$$

As usual

$$I(\epsilon) = \int \int_D f(x, y, z, U, U_x, U_y, U_z) dx dy dz.$$

The extremum is reached when:

$$\frac{\partial I}{\partial \epsilon}\bigg|_{\epsilon=0} = \int \int_D \left(\frac{\partial f}{\partial u}\eta + \frac{\partial f}{\partial u_x}\eta_x + \frac{\partial f}{\partial u_y}\eta_y + \frac{\partial f}{\partial u_z}\eta_z \right) dx dy dz = 0.$$

Applying Green's identity for the last three terms and a considerable amount of algebra produces the Euler-Lagrange differential equation for this case

$$\frac{\partial f}{\partial u} - \frac{\partial}{\partial x}\frac{\partial f}{\partial u_x} - \frac{\partial}{\partial y}\frac{\partial f}{\partial u_y} - \frac{\partial}{\partial z}\frac{\partial f}{\partial u_z} = 0.$$

An even more practical three-variable case, important in engineering dynamics, is when the Euclidean spatial coordinates are extended with time. Let us consider the variational problem of one temporal and two spatial dimensions as

$$I(u) = \int_{t_0}^{t_1} \int \int_D f(x, y, t, u, u_x, u_y, u_t) dx dy dt = \text{extremum}.$$

Here again

$$u_x = \frac{\partial u}{\partial x}; u_y = \frac{\partial u}{\partial y},$$

and

$$u_t = \frac{\partial u}{\partial t}.$$

We introduce the alternative solution as

$$U(x, y, t) = u(x, y, t) + \epsilon \eta(x, y, t),$$

with the temporal boundary conditions of

$$\eta(x, y, t_0) = \eta(x, y, t_1) = 0.$$

As above

$$I(\epsilon) = \int_{t_0}^{t_1} \int \int_D f(x, y, t, U, U_x, U_y, U_t) dx dy dt,$$

and the extremum is reached when:

$$\frac{\partial I}{\partial \epsilon}\Big|_{\epsilon=0} = \int_{t_0}^{t_1} \int \int_D \left(\frac{\partial f}{\partial u}\eta + \frac{\partial f}{\partial u_x}\eta_x + \frac{\partial f}{\partial u_y}\eta_y + \frac{\partial f}{\partial u_t}\eta_t \right) dx dy dt = 0. \quad (3.1)$$

The last member of the integral may be written as

$$\int_{t_0}^{t_1} \int \int_D \frac{\partial f}{\partial u_t}\eta_t dx dy dt = \int \int_D \int_{t_0}^{t_1} \frac{\partial f}{\partial u_t}\eta_t dt dx dy.$$

Integrating by parts yields

$$\int \int_D \left(\frac{\partial f}{\partial u_t}\eta \Big|_{t_0}^{t_1} - \int_{t_0}^{t_1} \eta \frac{\partial}{\partial t}\left(\frac{\partial f}{\partial u_t} \right) dt \right) dx dy.$$

Due to the temporal boundary condition, the first term vanishes and

$$-\int_{t_0}^{t_1} \int \int_D \eta \frac{\partial}{\partial t}\left(\frac{\partial f}{\partial u_t} \right) dx dy dt$$

remains. The second and third terms of Equation (3.1) may be rewritten by Green's identity as follows:

$$\int_{t_0}^{t_1} \cdot \int \int_D \left(\frac{\partial f}{\partial u_x} \eta_x + \frac{\partial f}{\partial u_y} \eta_y \right) dx dy dt =$$

$$- \int_{t_0}^{t_1} \int \int_D \eta \left(\frac{\partial}{\partial x} \left(\frac{\partial f}{\partial u_x} \right) + \frac{\partial}{\partial y} \left(\frac{\partial f}{\partial u_y} \right) \right) dx dy dt +$$

$$\int_{t_0}^{t_1} \int_{\partial D} \eta \left(\frac{\partial f}{\partial u_x} \frac{dy}{ds} + \frac{\partial f}{\partial u_y} \frac{dx}{ds} \right) ds dt.$$

With these changes, Equation (3.1) becomes

$$\frac{\partial I}{\partial \epsilon} \Big|_{\epsilon=0} = \int_{t_0}^{t_1} (\int \int_D \eta \left(\frac{\partial f}{\partial u} - \frac{\partial}{\partial x} \left(\frac{\partial f}{\partial u_x} \right) - \frac{\partial}{\partial y} \left(\frac{\partial f}{\partial u_y} \right) - \frac{\partial}{\partial t} \left(\frac{\partial f}{\partial u_t} \right) \right) dx dy +$$

$$\int_{\partial D} \eta \left(\frac{\partial f}{\partial u_x} \frac{dy}{ds} - \frac{\partial}{\partial u_y} \frac{dx}{ds} \right) ds) dt = 0.$$

Since the auxiliary function η is arbitrary, by the fundamental lemma of calculus of variations the first integral is only zero when

$$\frac{\partial f}{\partial u} - \frac{\partial}{\partial x} \left(\frac{\partial f}{\partial u_x} \right) - \frac{\partial}{\partial y} \left(\frac{\partial f}{\partial u_y} \right) - \frac{\partial}{\partial t} \left(\frac{\partial f}{\partial u_t} \right) = 0$$

in the interior of the domain D. This is the Euler-Lagrange differential equation of the problem. Since the boundary conditions of the auxiliary function were only temporal, the second integral is only zero when

$$\frac{\partial f}{\partial u_x} \frac{dy}{ds} - \frac{\partial}{\partial u_y} \frac{dx}{ds} = 0$$

on the boundary ∂D. This is the constraint of the variational problem. This result will be utilized in Chapter 11 in the solution of the elastic membrane vibration problem.

This may be generalized to four independent variables by using three spatial (x, y, z) and one temporal (t) variable. In those cases, the Euler-Lagrange differential equation is extended by an additional term:

$$\frac{\partial f}{\partial u} - \frac{\partial}{\partial x} \left(\frac{\partial f}{\partial u_x} \right) - \frac{\partial}{\partial y} \left(\frac{\partial f}{\partial u_y} \right) - \frac{\partial}{\partial y} \left(\frac{\partial f}{\partial u_z} \right) - \frac{\partial}{\partial t} \left(\frac{\partial f}{\partial u_t} \right) = 0$$

Such scenario will be instrumental in modeling continuum problems in Chapters 11 and 12.

3.6 Exercises

Find the Euler-Lagrange differential equation for the functionals.

1.
$$I = \int \int \left(x^2 \left(\frac{\partial u}{\partial x} \right)^2 + y^2 \left(\frac{\partial u}{\partial y} \right)^2 \right) dx dy = \text{extremum}.$$

2.
$$I = \int \int \left(\left(\frac{\partial u}{\partial t} \right)^2 - c^2 \left(\frac{\partial u}{\partial x} \right)^2 \right) dx dt = \text{extremum}.$$

3.
$$I = \int \int \left(\left(\frac{\partial u}{\partial x} \right)^2 + \left(\frac{\partial u}{\partial y} \right)^2 \right) dx dy = \text{extremum}.$$

4.
$$I = \int \int \left(\left(\frac{\partial u}{\partial x} \right)^2 + \left(\frac{\partial u}{\partial y} \right)^2 + \left(\frac{\partial u}{\partial z} \right)^2 \right) dx dy dz = \text{extremum}.$$
Constraint: $\int \int \int u^2 dx dy dz = 1$.

5.
$$I = \int \int \left(\frac{1}{2} \left(\frac{\partial z}{\partial x} \right)^2 + \frac{1}{2} \left(\frac{\partial z}{\partial y} \right)^2 + z \right) dx dy = \text{extremum}.$$

6.
$$I = \int_a^b (xy^2 + x^2 y + \dot{x}\dot{y}) dt = \text{extremum}.$$

Find the general solution for the following problems.

7.
$$I = \int (xy + \dot{x}\dot{y}) dt = \text{extremum}.$$

8.
$$I = \int \left(x^2 + y^2 + z^2 + (\dot{x})^2 + (\dot{y})^2 + (\dot{z})^2 \right) dt = \text{extremum}.$$

Find the Euler-Lagrange differential equation for the functionals.

9.
$$I = \int \int \sqrt{1 + \left(\frac{\partial u}{\partial x} \right)^2 + \left(\frac{\partial u}{\partial y} \right)^2} dx dy = \text{extremum}.$$

10.
$$I = \int \int \sqrt{1 + \left(\frac{\partial u}{\partial x} \right)^2 + \left(\frac{\partial u}{\partial y} \right)^2 + \left(\frac{\partial u}{\partial z} \right)^2} dx dy dz = \text{extremum}.$$

4

Higher order derivatives

The fundamental problem of calculus of variations involved the first derivative of the unknown function. In this chapter, we will allow the presence of higher order derivatives that lead to the so-called Euler-Poisson equation. The chapter will also present a method applying an algebraic constraint on the derivative. Finally the technique of linearization of second order problems will be discussed and illustrated.

4.1   The Euler–Poisson equation

First let us consider the variational problem of a functional with a single function, but containing its higher derivatives:

$$I(y) = \int_{x_0}^{x_1} f(x, y, y', \dots, y^{(m)}) dx.$$

Accordingly, boundary conditions for all derivatives will also be given as

$$y(x_0) = y_0, y(x_1) = y_1,$$

$$y'(x_0) = y'_0, y'(x_1) = y'_1,$$

$$y''(x_0) = y''_0, y''(x_1) = y''_1,$$

and so on until

$$y^{(m-1)}(x_0) = y_0^{(m-1)}, y^{(m-1)}(x_1) = y_1^{(m-1)}.$$

As in the past chapters, we introduce an alternative solution of

$$Y(x) = y(x) + \epsilon\eta(x),$$

where the arbitrary auxiliary function $\eta(x)$ is continuously differentiable on the interval $x_0 \leq x \leq x_1$ and satisfies

$$\eta(x_0) = 0, \eta(x_1) = 0.$$

The variational problem in terms of the alternative solution is

$$I(\epsilon) = \int_{x_0}^{x_1} f(x, Y, Y', \ldots, Y^{(m)}) dx.$$

The differentiation with respect to ϵ follows

$$\frac{dI}{d\epsilon} = \int_{x_0}^{x_1} \frac{d}{d\epsilon} f(x, Y, Y', \ldots, Y^{(m)}) dx,$$

and by using the chain rule, the integrand is reshaped as

$$\frac{\partial f}{\partial Y} \frac{dY}{d\epsilon} + \frac{\partial f}{\partial Y'} \frac{dY'}{d\epsilon} + \frac{\partial f}{\partial Y''} \frac{dY''}{d\epsilon} + \ldots + \frac{\partial f}{\partial Y^{(m)}} \frac{dY^{(m)}}{d\epsilon}.$$

Substituting the alternative solution and its derivatives with respect to ϵ, the integrand yields

$$\frac{\partial f}{\partial Y} \eta + \frac{\partial f}{\partial Y'} \eta' + \frac{\partial f}{\partial Y''} \eta'' + \ldots + \frac{\partial f}{\partial Y^{(m)}} \eta^{(m)}.$$

Hence the functional becomes

$$\frac{dI}{d\epsilon} = \int_{x_0}^{x_1} \left(\frac{\partial f}{\partial Y} \eta + \frac{\partial f}{\partial Y'} \eta' + \frac{\partial f}{\partial Y''} \eta'' + \ldots + \frac{\partial f}{\partial Y^{(m)}} \eta^{(m)} \right) dx.$$

Integrating by term results in

$$\frac{dI}{d\epsilon} = \int_{x_0}^{x_1} \frac{\partial f}{\partial Y} \eta \, dx + \int_{x_0}^{x_1} \frac{\partial f}{\partial Y'} \eta' \, dx + \int_{x_0}^{x_1} \frac{\partial f}{\partial Y''} \eta'' \, dx + \ldots + \int_{x_0}^{x_1} \frac{\partial f}{\partial Y^{(m)}} \eta^{(m)} \, dx,$$

and integrating by parts produces

$$\frac{dI}{d\epsilon} = \int_{x_0}^{x_1} \eta \frac{\partial f}{\partial Y} dx - \int_{x_0}^{x_1} \eta \frac{d}{dx} \frac{\partial f}{\partial Y'} dx + \int_{x_0}^{x_1} \eta \frac{d^2}{dx^2} \frac{\partial f}{\partial Y''} dx -$$

$$\ldots (-1)^m \int_{x_0}^{x_1} \eta \frac{d^{(m)}}{dx^{(m)}} \frac{\partial f}{\partial Y^{(m)}} dx.$$

Factoring the auxiliary function and combining the terms again simplifies to

$$\frac{dI}{d\epsilon} = \int_{x_0}^{x_1} \eta \left(\frac{\partial f}{\partial Y} - \frac{d}{dx} \frac{\partial f}{\partial Y'} + \frac{d^2}{dx^2} \frac{\partial f}{\partial Y''} - \ldots (-1)^m \frac{d^{(m)}}{dx^{(m)}} \frac{\partial f}{\partial Y^{(m)}} \right) dx.$$

Finally the extremum at $\epsilon = 0$ and the fundamental lemma produces the **Euler-Poisson equation**

$$\frac{\partial f}{\partial y} - \frac{d}{dx} \frac{\partial f}{\partial y'} + \frac{d^2}{dx^2} \frac{\partial f}{\partial y''} - \ldots (-1)^m \frac{d^{(m)}}{dx^{(m)}} \frac{\partial f}{\partial y^{(m)}} = 0.$$

The Euler-Poisson equation is an ordinary differential equation of order $2m$ and requires the aforementioned $2m$ boundary conditions, where m is the highest order derivative contained in the functional.

In reality, the highest order most frequently used is $m = 2$ reflecting the fact that many natural phenomena are described by second order differential equations. In geometry the curvature, and in physics the acceleration, is proportional to the second derivative, giving rise to many practical variational problems, some discussed in later sections.

To illustrate the Euler-Poisson equation solution, we consider the functional

$$I = \int_0^{\pi/4} (y''^2 - y^2 + x^2)dx = \text{extremum}.$$

The boundary conditions are

$$y(0) = 0, y(\pi/4) = \frac{1}{\sqrt{2}},$$

and

$$y'(0) = 1, y'(\pi/4) = \frac{1}{\sqrt{2}}.$$

The second order form of the Euler-Poisson equation is proper for this problem. The components are

$$\frac{\partial f}{\partial y} = -2y,$$

$$\frac{\partial f}{\partial y'} = 0,$$

and

$$\frac{\partial f}{\partial y''} = 2y''.$$

The differential equation becomes

$$-2y - 0 + \frac{d^2}{dx^2}(2y'') = 0,$$

which simplifies to

$$\frac{d^4 y}{dx^4} - y = 0.$$

This is incidentally the same fourth order differential equation obtained in the example in Section 3.1.1 as a problem of two functions. Borrowing from there, the solution form will be

$$y(x) = ae^x + be^{-x} + c\cos(x) + d\sin(x).$$

Applying the four boundary conditions, we obtain a different system of equations. Specifically, the boundary conditions here will include derivative conditions also. The derivative is

$$y'(x) = ae^x - be^{-x} - c\sin(x) + d\cos(x).$$

At the lower end, the displacement condition results in

$$y(0) = ae^0 + be^0 + c\cos(0) + d\sin(0) = a + b + c = 0.$$

The first derivative condition brings

$$y'(0) = ae^0 - be^0 - c\sin(0) + d\cos(0) = a - b + d = 1.$$

Similarly at the upper end

$$y(\frac{\pi}{4}) = ae^{\frac{\pi}{4}} + be^{\frac{-\pi}{4}} + c\cos(\frac{\pi}{4}) + d\sin(\frac{\pi}{4}) = ae^{\frac{\pi}{4}} + be^{\frac{-\pi}{4}} + \frac{c}{\sqrt{2}} + \frac{d}{\sqrt{2}} = \frac{1}{\sqrt{2}}$$

and

$$y'(\frac{\pi}{4}) = ae^{\frac{\pi}{4}} - be^{\frac{-\pi}{4}} - c\sin(\frac{\pi}{4}) + d\cos(\frac{\pi}{4}) = ae^{\frac{\pi}{4}} - be^{\frac{-\pi}{2}} - \frac{c}{\sqrt{2}} + \frac{d}{\sqrt{2}} = \frac{1}{\sqrt{2}}.$$

Without details of the elementary algebra solving the system of four equations, we conclude

$$a = 0, b = 0, c = 0, d = 1.$$

Hence the solution to the problem is

$$y = \sin(t).$$

The similarities between the two problems belonging to a different class are important as they illustrate the possibility of solving problems with alternative methods.

4.2   The Euler–Poisson system of equations

In the case of a functional with multiple functions along with their higher order derivatives, the problem gets more difficult. Assuming p functions in the functional, the problem is posed in the form of

$$I(y_1, \ldots, y_p) = \int_{x_0}^{x_1} f(x, y_1, y_1', \ldots, y_1^{(m_1)}, \ldots, y_p, y_p', \ldots, y_p^{(m_p)}) dx.$$

Note that the highest order of the derivative of the various functions is not necessarily the same. This is a rather straightforward generalization of the case of the last section, leading to a system of Euler-Poisson equations as follows:

$$\frac{\partial f}{\partial y_1} - \frac{d}{dx}\frac{\partial f}{\partial y_1'} + \frac{d^2}{dx^2}\frac{\partial f}{\partial y_1''} - \cdots (-1)^{m_1}\frac{d^{(m_1)}}{dx^{(m_1)}}\frac{\partial f}{\partial y_1^{(m_1)}} = 0,$$

$$\cdots,$$

$$\frac{\partial f}{\partial y_p} - \frac{d}{dx}\frac{\partial f}{\partial y_p'} + \frac{d^2}{dx^2}\frac{\partial f}{\partial y_p''} - \cdots (-1)^{m_p}\frac{d^{(m_p)}}{dx^{(m_p)}}\frac{\partial f}{\partial y_p^{(m_p)}} = 0.$$

This is a set of p ordinary differential equations that may or may not be coupled, hence resulting in a varying level of difficulty in solution.

As mentioned before, the highest orders are mostly second order, and two equations are used describing a pair of functions. Such scenario occurs in geometrical problems involving the curvature of the curves presented in parametric form.

We illustrate the use of a system of Euler-Poisson equations with the following variational problem.

$$I = \int_{t=0}^{t=1} ((\ddot{x})^2 + (\ddot{y})^2)dt = \text{extremum}.$$

The equations will be

$$\frac{\partial f}{\partial x} \quad \frac{d}{dt}\frac{\partial f}{\partial \dot{x}} + \frac{d^2}{dt^2}\frac{\partial f}{\partial \ddot{x}} = \frac{d^2}{dt^2}2\ddot{x} = 0,$$

and

$$\frac{\partial f}{\partial y} - \frac{d}{dt}\frac{\partial f}{\partial \dot{y}} + \frac{d^2}{dt^2}\frac{\partial f}{\partial \ddot{y}} = \frac{d^2}{dt^2}2\ddot{y} = 0.$$

The system is decoupled

$$\frac{d^4}{dt^4}x(t) = 0,$$

and

$$\frac{d^4}{dt^4}y(t) = 0.$$

Hence, the solutions will be simply

$$x = a_0 + a_1 t + a_2 t^2 + a_3 t^3,$$

and

$$y = b_0 + b_1 t + b_2 t^2 + b_3 t^3.$$

The constants are resolved by different boundary conditions for the two functions albeit for both with a pair of displacement and derivative conditions. For the $x(t)$ function

$$x(0) = 0, x(1) = 1, \dot{x}(0) = 0, \dot{x}(1) = 1,$$

yields

$$x(0) = 0 \rightarrow a_0 = 0,$$

and

$$\dot{x}(0) = 0 \rightarrow a_1 = 0.$$

For the remaining coefficients, we get the system of

$$x(1) = 1 = a_2 + a_3,$$

and

$$\dot{x}(1) = 1 = 2a_2 + 3a_3.$$

The solution is

$$a_2 = 2, a_3 = -1,$$

hence, the solution of the first function is

$$x(t) = 2t^2 - t^3.$$

Applying different boundary conditions for the second function

$$y(0) = 0, y(1) = 1, \dot{y}(0) = 2, \dot{y}(1) = 2,$$

produces

$$y(t) = 2t - 3t^2 + 2t^3.$$

The solution curve is a parametric spline function shown in Figure 4.1, a subject of deeper discussion in the computational geometry chapter.

Since the ratio of the parametric function second derivatives is related to the explicit function second derivatives, in essence the problem was to find a minimal curvature solution to the functional posted. It is very visible that the goal of the optimization was achieved, the spline function traversing the span is certainly smooth.

It is easy to verify that the location boundary conditions are simultaneously satisfied. Considering the derivative conditions, computing

$$\frac{dy}{dx} = \frac{\dot{y}}{\dot{x}}$$

would result in infinity at the lower end which is demonstrated by the vertical tangent of the solution there. At the upper end, the tangent adheres to the computed value of 2.

The system of Euler-Poisson equations may be coupled resulting in a difficult solution of the participating differential equations. Third order applications with three equations may also occur in geometrical problems in a 3-dimensional space where the torsion of the parametrically represented curve is also used.

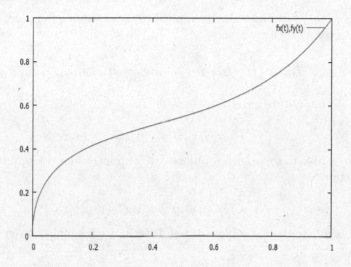

FIGURE 4.1 Parametric solution

4.3 Algebraic constraints on the derivative

It is also common in engineering applications to impose constraint conditions on some of the derivatives participating in the functional, as in the problem

$$I(y) = \int_{x_0}^{x_1} f(x, y, y')dx = \text{extremum},$$

subject to

$$g(x, y, y') = 0.$$

In order to be able to solve such problems, we need to introduce a Lagrange multiplier as a function of the independent variable as

$$h(x, y, y', \lambda) = f(x, y, y') + \lambda(x)g(x, y, y').$$

The use of this approach means that the functional now contains two unknown functions and the variational problem becomes

$$I(y, \lambda) = \int_{x_0}^{x_1} h(x, y, y', \lambda)dx,$$

with the original boundary conditions, but without a constraint. The solution is obtained for the two function case by a system of two Euler-Lagrange equations.

Derivative constraints may also be applied to the case of higher order derivatives. The second order problem of

$$I(y) = \int_{x_0}^{x_1} f(x, y, y', y'')dx = \text{extremum}$$

may be subject to a constraint

$$g(x, y, y', y'') = 0.$$

In order to be able to solve such problems, we also introduce a Lagrange multiplier function as

$$h(x, y, y', y'') = f(x, y, y', y'') + \lambda(x)g(x, y, y', y'').$$

The result is a variational problem of two functions with higher order derivatives as

$$I(y, \lambda) = \int_{x_0}^{x_1} h(x, y, y', y'', \lambda)dx = \text{extremum}.$$

Hence, the solution may be obtained by the application of a system of two Euler-Poisson equations.

Furthermore, derivative constraints may also be applied to a variational problem originally exhibiting multiple functions, such as

$$I(y, z) = \int_{x_0}^{x_1} f(x, y, y', z, z')dx = \text{extremum}$$

subject to

$$g(x, y, y', z, z') = 0.$$

Here the new functional is

$$h(x, y, y', z, z', \lambda) = f(x, y, y', z, z') + \lambda(x)g(x, y, y', z, z').$$

Following above, this problem translates into the unconstrained form of

$$I(y, z, \lambda) = \int_{x_0}^{x_1} h(x, y, y', z, z', \lambda)dx$$

that may be solved by a system of three Euler-Lagrange differential equations

$$\frac{\partial h}{\partial y} - \frac{d}{dx}\frac{\partial h}{\partial y'} = 0,$$

$$\frac{\partial h}{\partial z} - \frac{d}{dx}\frac{\partial h}{\partial z'} = 0,$$

and

$$\frac{\partial h}{\partial \lambda} - \frac{d}{dx}\frac{\partial h}{\partial \lambda'} = 0.$$

For illustration, we consider the variational problem of

$$I(y, z) = \int_{x_0}^{x_1} (y^2 - z^2)dx = \text{extremum},$$

under the derivative constraint of

$$y' + y - z = 0.$$

This results in

$$h(x, y, y', z, z', \lambda) = y^2 - z^2 + \lambda(x)(y' - y + z).$$

The solution is obtained from the following three equations

$$2y - \lambda - \lambda' = 0,$$

$$-2z + \lambda = 0,$$

and

$$y' + y - z = 0.$$

The second equation enables the elimination of the Lagrange multiplier, resulting in the linear system of first order differential equations

$$y - z - z' = 0,$$

and

$$y' + y - z = 0.$$

This may be recast in a matrix form as

$$\begin{bmatrix} z' \\ y' \end{bmatrix} = \begin{bmatrix} -1 & 1 \\ 1 & -1 \end{bmatrix} \begin{bmatrix} z \\ y \end{bmatrix}.$$

Computing the matrix exponential or the fundamental solution matrix based on the characteristic equation solution, the system may easily be solved.

4.4   Linearization of second order problems

It is very common in engineering practice that the highest derivative of interest is of second order. As mentioned before, accelerations in engineering analysis of motion, and other important application concepts are tied to the second derivative.

This specific case of second order problems may be reverted to a linear problem involving two functions. Consider

$$I(y) = \int_{x_0}^{x_1} f(x, y, y', y'') dx = \text{extremum}$$

with the following boundary conditions

$$y(x_0) = y_0, y(x_1) = y_1, y'(x_0) = y_0', y'(x_1) = y_1'.$$

By introducing a new function

$$z(x) = y'(x),$$

we can reformulate the unconstrained second order variational problem as a constrained variational problem of the first order with multiple functions:

$$I(y, z) = \int_{x_0}^{x_1} f(x, y, z, z') dx = \text{extremum},$$

subject to a constraint involving the derivative

$$g(x, y, z) = z - y' = 0.$$

Using the combined function in the form of

$$h(x, y, z, z', \lambda) = f(x, y, z, z') + \lambda(x)(z - y'),$$

and following the process laid out in the last section, we can produce a system of three Euler-Lagrange differential equations:

$$\frac{\partial h}{\partial y} - \frac{d}{dx}\frac{\partial h}{\partial y'} = \frac{\partial f}{\partial y} + \frac{d\lambda}{dx} = 0,$$

$$\frac{\partial h}{\partial z} - \frac{d}{dx}\frac{\partial h}{\partial z'} = \frac{\partial f}{\partial z} + \lambda - \frac{d}{dx}\frac{\partial f}{\partial z'} = 0,$$

and

$$\frac{\partial h}{\partial \lambda} - \frac{d}{dx}\frac{\partial h}{\partial \lambda'} = z - y' = 0.$$

This may, of course, be turned into the Euler-Poisson equation by expressing

$$\lambda = \frac{d}{dx}\frac{\partial f}{\partial z'} - \frac{\partial f}{\partial z}$$

from the middle equation and differentiating as

$$\frac{d\lambda}{dx} = \frac{d^2}{dx^2}\frac{\partial f}{\partial z'} - \frac{d}{dx}\frac{\partial f}{\partial z}.$$

Substituting this and the third equation into the first yields the Euler-Poisson equation we could have achieved, had we approached the original quadratic problem directly:

$$\frac{\partial f}{\partial y} - \frac{d}{dx}\frac{\partial f}{\partial y'} + \frac{d^2}{dx^2}\frac{\partial f}{\partial y''} = 0.$$

Depending on the particular application circumstances, however, the linear system of Euler-Lagrange equations may be more conveniently solved than the quadratic Euler-Poisson equation.

We illustrate this process by the following problem:

$$\int (y^2 - y''^2)dx = \text{extremum}.$$

We use

$$z = y', z' = y'',$$

to substitute and produce the constrained but linear problem

$$\int \left(y^2 - z'^2 + \lambda(z - y')\right) dx = \text{extremum}.$$

The Euler-Poisson system of equations becomes

$$\frac{\partial f}{\partial y} + \frac{d\lambda}{dx} = 2y + \lambda' = 0,$$

$$\frac{\partial f}{\partial z} + \lambda - \frac{d}{dx}\frac{\partial f}{\partial z'} = \lambda + 2z' = 0,$$

and the third equation is simply the dictated substitution

$$z - y' = 0.$$

In this case, it is not possible to eliminate one of the equations since all three variables have derivatives also. Hence, the linear system of first order differential equations becomes

$$\lambda' = -2y,$$

$$z' = -\lambda/2$$

and

$$y' = z.$$

Recasting it in matrix form results in

$$\begin{bmatrix} \lambda' \\ z' \\ y' \end{bmatrix} = \begin{bmatrix} 0 & 0 & -2 \\ -1/2 & 0 & 0 \\ 0 & 1 & 0 \end{bmatrix} \begin{bmatrix} \lambda \\ z \\ y \end{bmatrix}.$$

The solution of this system would lead to the same result as would be obtained by attacking the original problem with the Euler-Poisson equation approach. While it appears that solving the latter form would be more direct, it depends on the particular example and is not always the case.

Finally, practical applications of higher than second order derivatives also do occur albeit not very frequently. Computer aided manufacturing desires tool paths that are smooth space curves not just with minimal curvature, but also torsion, bringing in the need for the third derivative.

4.5   Exercises

Find the solutions for the variational problems.

1.
$I = \int_0^1 (yy' + y''^2)dx = \text{extremum}.$
Boundary conditions: $y(0) = 0, y'(0) = 1, y(1) = 2, y'(1) = 4.$

2.
$I = \int_0^\infty (y^2 + y'^2 + (y'' + y')^2)dx = \text{extremum}.$
Boundary conditions: $y(0) = 1, y'(0) = 2, y(\infty) = 0, y'(\infty) = 0.$

3.
$I = \int_0^1 (1 + y''^2)dx = \text{extremum}.$
Boundary conditions: $y(0) = 0, y'(0) = 1, y(1) = 1, y'(1) = 1.$

4.
$I = \int_a^b (y^2 + 2y'^2 + y''^2)dx = \text{extremum}.$

5.
$I = \int_0^1 y''^2 dx = \text{extremum}.$
Boundary conditions: $y(0) = 0, y'(0) = \frac{1}{2}, y(1) = 1, y'(1) = \frac{3}{2}.$

6.
$I = \int_0^\pi (y''^2 - 4y^2)dx = \text{extremum}.$

5

The inverse problem

This chapter deals with the case when the engineer starts from a differential equation with certain boundary conditions that is difficult to solve. Executing the inverse of the Euler-Lagrange process and obtaining the variational formulation of the boundary value problem may be advantageous. This discussion leads to the variational form of eigenvalue problems and the Sturm-Liouville class of differential equations. The latter are the source of many notable orthogonal polynomial families, such as Legendre's, which will be presented in detail.

5.1 Linear differential operators

It is not necessarily easy, or may not even be possible, to reconstruct the variational problem from a differential equation. For differential equations, partial or ordinary, containing a linear, self-adjoint, positive operator, the task may be accomplished. Such an operator exhibits the relation

$$(Au, v) = (u, Av),$$

where the parenthesis expression denotes a scalar product in the function space of the solution of the differential equation. Positive definiteness of the operator means

$$(Au, u) \geq 0,$$

with zero attained only for the trivial ($u = 0$) solution. Let us consider the differential equation of

$$Au = f,$$

where the operator obeys the above conditions and f is a known function. If the differential equation has a solution, it corresponds to the minimum value of the functional

$$I(u) = \frac{1}{2}(Au, u) - (u, f).$$

This may be proven by simply applying the appropriate Euler-Lagrange equation to this functional.

5.2   The variational form of Poisson's equation

We demonstrate the inverse process through the example of Poisson's equation, a topic of much interest for engineers:

$$\Delta u(x) = \frac{\partial^2 u}{\partial x^2} + \frac{\partial^2 u}{\partial y^2} = f(x, y).$$

Here the left-hand side is the Laplace operator which fulfills the above requirements on the operator. We impose Dirichlet type boundary conditions on the boundary of the domain of interest.

$$u(x, y) = 0; (x, y) \in \partial D,$$

where D is the domain of solution and ∂D is its boundary. According to the above proposition, we need to compute

$$(Au, u) = \int \int_D u \left(\frac{\partial^2 u}{\partial x^2} + \frac{\partial^2 u}{\partial y^2} \right) dx dy.$$

Applying Green's theorem results in

$$(Au, u) = -\int_{\partial D} \left(u \frac{\partial u}{\partial y} dx - u \frac{\partial u}{\partial x} dy \right) - \int \int_D \left(\frac{\partial u}{\partial x} \right)^2 + \left(\frac{\partial u}{\partial y} \right)^2 dx dy.$$

Due to the boundary conditions, the first term vanishes and we obtain

$$(Au, u) = -\int \int_D \left(\frac{\partial u}{\partial x} \right)^2 + \left(\frac{\partial u}{\partial y} \right)^2 dx dy.$$

The right-hand side term of the differential equation is processed as

$$(u, f) = \int \int_D u f(x, y) dx dy.$$

The variational formulation of Poisson's equation finally is

$$\int \int_D \left(-\frac{1}{2} \left(\left(\frac{\partial u}{\partial x} \right)^2 + \left(\frac{\partial u}{\partial y} \right)^2 \right) - u f \right) dx dy = \int \int_D F dx dy = \text{extremum}.$$

To prove this, we will apply the Euler-Lagrange equation developed in Section 3.3. The components for this particular case are:

$$\frac{\partial F}{\partial u} = -f,$$

$$\frac{\partial}{\partial x}\frac{\partial F}{\partial u_x} = -\frac{\partial}{\partial x}u_x = -\frac{\partial^2 u}{\partial x^2},$$

and

$$\frac{\partial}{\partial y}\frac{\partial F}{\partial u_y} = -\frac{\partial}{\partial y}u_y = -\frac{\partial^2 u}{\partial y^2}.$$

The Euler-Lagrange differential equation of

$$-f + \frac{\partial^2 u}{\partial x^2} + \frac{\partial^2 u}{\partial y^2} = 0$$

is clearly equivalent with Poisson's equation after ordering.

5.3 The variational form of eigenvalue problems

Eigenvalue problems of various kinds may also be formulated as variational problems [8]. We consider the equation of the form

$$\Delta u(x, y) - \lambda u(x, y) = 0, \tag{5.1}$$

where the unknown function $u(x, y)$ defined on domain D is the eigenfunction and λ is the eigenvalue. The boundary condition is imposed as

$$u(x, y) = 0$$

on the perimeter ∂D of the domain D. The corresponding variational problem is of the form

$$I = \int\int_D \left(\left(\frac{\partial u}{\partial x}\right)^2 + \left(\frac{\partial u}{\partial y}\right)^2\right) dx dy = \text{extremum}, \tag{5.2}$$

under the normalization condition of

$$g(x, y) = \int\int_D u^2(x, y) dx dy = 1.$$

This relation is proven as follows. Following the Lagrange solution of constrained variational problems introduced in Section 2.2, we can write

$$h(x, y) = u(x, y) + \lambda g(x, y),$$

and

$$I = \int\int_D \left(\left(\frac{\partial u}{\partial x}\right)^2 + \left(\frac{\partial u}{\partial y}\right)^2 + \lambda u^2\right) dx dy.$$

Note that the λ is both in the role of the Lagrange multiplier and the eigenvalue. Introducing

$$U(x,y) = u(x,y) + \epsilon\eta(x,y)$$

the variational form becomes

$$I(\epsilon) = \int\int_D \left(\left(\frac{\partial u}{\partial x} + \epsilon\frac{\partial\eta}{\partial x}\right)^2 + \left(\frac{\partial u}{\partial y} + \epsilon\frac{\partial\eta}{\partial y}\right)^2 + \lambda(u + \epsilon\eta)^2 \right) dxdy.$$

The extremum is reached when

$$\frac{dI(\epsilon)}{d\epsilon}\bigg|_{\epsilon=0} = 0,$$

which gives rise to the equation

$$2\int\int_D \left(\frac{\partial u}{\partial x}\frac{\partial\eta}{\partial x} + \frac{\partial u}{\partial y}\frac{\partial\eta}{\partial y} + \lambda u\eta \right) dxdy = 0. \tag{5.3}$$

Green's identity in its original three-dimensional form was exploited on several occasions earlier; here we apply it for the special vector field

$$\eta\nabla u$$

in a two-dimensional domain. The result is

$$\int\int_D (\nabla\eta \cdot \nabla u)dA = \int_{\partial D} \eta(\nabla u \cdot \underline{n})ds - \int\int_D \eta\nabla^2 u dA.$$

Since the tangent of the circumference is in the direction of

$$dx\,\underline{i} + dy\,\underline{j},$$

the unit normal may be computed as

$$\underline{n} = \frac{dy\,\underline{i} - dx\,\underline{j}}{\sqrt{dx^2 + dy^2}}.$$

Finally utilizing the arc length formula of

$$ds = \sqrt{dx^2 + dy^2},$$

the line integral over the circumference of the domain becomes

$$\int_{\partial D} \eta\left(\frac{\partial u}{\partial x}dy - \frac{\partial u}{\partial y}dx\right).$$

Applying above for the first two terms of Equation (5.3) results in

$$\int\int_D \left(\frac{\partial u}{\partial x}\frac{\partial\eta}{\partial x} + \frac{\partial u}{\partial y}\frac{\partial\eta}{\partial y}\right) dxdy =$$

$$\iint_{\partial D} \left(\eta \left(\frac{\partial u}{\partial x} \underline{i} + \frac{\partial u}{\partial y} \underline{j} \right) \cdot \underline{n} \right) ds - \iint_D \left(\frac{\partial^2 u}{\partial x^2} + \frac{\partial^2 u}{\partial y^2} \right) \eta dx dy =$$

$$\int_{\partial D} \frac{\partial u}{\partial x} \eta dy - \frac{\partial u}{\partial y} \eta dx - \iint_D \Delta u \eta dx dy.$$

The integral over the boundary vanishes due to the assumption on η and substituting the remainder part into Equation (5.3) we obtain

$$-2 \iint_D (\Delta u - \lambda u) \eta dx dy = 0.$$

Since $\eta(x, y)$ is arbitrarily chosen, in order to satisfy this equation

$$\Delta u - \lambda u = 0$$

must be satisfied. Thus we have established that Equation (5.2) is indeed the variational form of Equation (5.1) and the Lagrange multiplier is the eigenvalue.

5.3.1 Orthogonal eigensolutions

The eigenvalue problem has an infinite number of eigenvalues, and for each eigenvalue there exists a corresponding eigenfunction that is unique apart from a constant factor. Hence, the variational form should also provide means for the solution of multiple pairs.

Let us denote the series of eigenpairs as

$$(\lambda_1, u_1), (\lambda_2, u_2), \dots (\lambda_n, u_n).$$

Assuming that we have already found the first pair satisfying

$$\Delta u_1 - \lambda_1 u_1 = 0,$$

we seek the second solution $u_2, \lambda_2 \neq \lambda_1$ following the process laid out in the last section. Then for any arbitrary auxiliary function η it follows that

$$\iint_D \left(\frac{\partial u_2}{\partial x} \frac{\partial \eta}{\partial x} + \frac{\partial u_2}{\partial y} \frac{\partial \eta}{\partial y} + \lambda_2 u_2 \eta \right) dx dy = 0.$$

Applying an auxiliary function of the special form of

$$\eta = u_1,$$

we obtain

$$\iint_D \left(\frac{\partial u_2}{\partial x} \frac{\partial u_1}{\partial x} + \frac{\partial u_2}{\partial y} \frac{\partial u_1}{\partial y} + \lambda_2 u_2 u_1 \right) dx dy = 0.$$

From the first eigenpair, we know

$$\int \int_D \left(\frac{\partial u_1}{\partial x} \frac{\partial \eta}{\partial x} + \frac{\partial u_1}{\partial y} \frac{\partial \eta}{\partial y} + \lambda_1 u_1 \eta \right) dxdy = 0.$$

Applying an auxiliary function of the special form of

$$\eta = u_2,$$

we obtain

$$\int \int_D \left(\frac{\partial u_1}{\partial x} \frac{\partial u_2}{\partial x} + \frac{\partial u_1}{\partial y} \frac{\partial u_2}{\partial y} + \lambda_1 u_1 u_2 \right) dxdy = 0.$$

Subtracting the equations and canceling the identical terms results in

$$(\lambda_2 - \lambda_1) \int \int_D u_1 u_2 dxdy = 0.$$

Since

$$\lambda_1 \neq \lambda_2,$$

it follows that

$$\int \int_D u_2 u_1 dxdy = 0$$

must be true. The two eigenfunctions are orthogonal. With similar arguments and specially selected auxiliary functions, it is also easy to show that the second solutions also satisfy

$$\Delta u_2 - \lambda_2 u_2 = 0.$$

The subsequent eigensolutions may be found by the same procedure and the sequence of the eigenpairs attain the extrema of the variational problem under the successive conditions of the orthogonality against the preceding solutions.

5.4   Sturm–Liouville problems

The process demonstrated in the last section in connection with Laplace's operator may be applied to arrive at eigenvalues and eigenfunctions of other differential equations as well. Introducing the Sturm-Liouville operator

$$L_{S-L} = \frac{d}{dx}\left(p(x)\frac{d}{dx}\right) + q(x),$$

where the functions $p(x) > 0, q(x)$, and $r(x) > 0$ are continuous and continuously differentiable, the eigenvalue problem of the form

$$L_{S-L}y(x) + \lambda r(x)y(x) = 0$$

is called a Sturm-Liouville eigenproblem. Here the unknown solution $y(x)$ is the eigenfunction and λ is the eigenvalue. The eigenvalues of these problems are all real, non-negative and form a strictly increasing sequence. For each eigenvalue, there exists one and only one linearly independent eigenfunction.

This corresponds to the variational problem of

$$I(y) = \int_{x_0}^{x_1} \left(p(x)y'^2 - q(x)y^2 \right) dx = \text{extremum},$$

subject to the constraint of

$$\int_{x_0}^{x_1} r(x)y^2(x)dx = 1.$$

The Lagrange multiplier constrained form is

$$I(y) = \int_{x_0}^{x_1} \left(p(x)y'^2 - q(x)y^2 - \lambda r(x)y^2 \right) dx = \text{extremum}.$$

The Euler-Lagrange differential equation becomes

$$-2q(x)y - 2\lambda r(x)y - \frac{d}{dx}2p(x)y' = 0,$$

or after shortening and ordering

$$\frac{d}{dx}p(x)y' + q(x)y + \lambda r(x)y = 0.$$

These are called the Sturm-Liouville differential equations. The boundary conditions imposed are either regular,

$$y(x_0) = y_0, y(x_1) = y_1,$$

or periodic

$$y(x_0) = y(x_1), y'(x_0) = y'(x_1).$$

Note that the same functions with different integration limits and boundary conditions produce a so-called Sturm-Liouville series that is not discussed further here.

To illustrate the Sturm-Liouville equations and their solutions, we consider the functions

$$p(x) = 1, q(x) = 0, r(x) = 1.$$

The equation for these functions is

$$\frac{d}{dx}y' + \lambda y(x) = 0.$$

Executing the differentiation, we obtain the equation

$$y'' + \lambda y = 0,$$

whose solutions

$$y_i(x) = c_i \sin(ix), i = 1, 2, \ldots$$

would satisfy the equation with a judicious choice of the constant c_i and boundary conditions. We will assume the regular boundary conditions of

$$y(0) = 0, y(\pi) = 0.$$

The selection of this boundary is for the convenience of dealing with this function family and does not restrict the generality of the discussion.

In order to establish the eigenvalues, we compute

$$\frac{d^2}{dx^2}\left(\sin(ix)\right) = \frac{d}{dx}\left(i\cos(ix)\right) = -i^2\sin(ix),$$

and substitute into the differential equation

$$-i^2 c_i \sin(ix) + \lambda c_i \sin(ix) = 0.$$

This is satisfied if the eigenvalue is

$$\lambda = i^2.$$

The constraint equation aids in recovering the constants. For $i = 1$

$$\int_0^\pi \left(c_1 \sin(1x)\right)^2 dx = c_1^2 \int_0^\pi \sin^2(x) dx = c_1^2 \left(\frac{1}{2}x\Big|_0^\pi - \frac{1}{4}\sin(2x)\Big|_0^\pi\right).$$

Substituting the boundary values results in

$$c_1^2 \frac{\pi}{2} = 1,$$

producing the constant necessary to satisfy the constraint:

$$c_1 = \sqrt{\frac{2}{\pi}}.$$

The generic coefficients may be obtained from the form

$$\int_0^\pi \left(c_i \sin(ix)\right)^2 dx = c_i^2 \int_0^\pi \sin^2(ix) dx = c_i^2 \left(\frac{1}{2}x\Big|_0^\pi - \frac{1}{4i}\sin(2ix)\Big|_0^\pi\right).$$

Considering the boundary conditions, the second term vanishes and

$$c_i^2 \frac{\pi}{2} = 1,$$

which yields the generic coefficients as

$$c_i = \sqrt{\frac{2}{\pi}}.$$

A more generic discussion of this case is presented in [7]. It is noteworthy that even this simplest form of Sturm-Liouville problems leads to an engineering application, the vibrating string problem, the subject of Section 11.1.

5.4.1 Legendre's equation and polynomials

A very important member of the Sturm-Liouville class of problems is obtained by the selection of

$$p(x) = 1 - x^2, q(x) = 0, r(x) = 1.$$

defined in the interval

$$(x_0, x_1) = (-1, 1).$$

The corresponding variational problem following the last section is of the form

$$I(y) = \int_{-1}^{1} \left((1 - x^2)(y')^2 - \lambda y^2\right) dx = \text{extremum}.$$

After applying the Euler-Lagrange differential equation, the specific Sturm-Liouville equation becomes

$$\frac{d}{dx}\left((1 - x^2)y'\right) + \lambda y(x) = 0.$$

Differentiation yields Legendre's differential equation

$$(1 - x^2)y'' - 2xy' + \lambda y = 0.$$

The solution around $x_0 = 0$ may be sought in the form of a power series as

$$y = \sum_{i=0}^{\infty} a_i x^i,$$

since $x_0 = 0$ is an ordinary point, albeit the boundary points are regular singular points.

Substituting this solution and its derivatives as needed yields the generic recurrence formula

$$a_{i+2} = \frac{i(i+1) - \lambda_i}{(i+2)(i+1)} a_i,$$

and the equation

$$(-2 + \lambda_1)a_1 x + \lambda_0 a_0 = 0.$$

This defines the first two eigenvalues as

$$\lambda_0 = 0, \lambda_1 = 2,$$

with yet undefined a_0, a_1 coefficients. This leads to a solution of two series

$$y(x) = a_0 + a_2 x^2 + a_4 x^4 + ... + a_1 x + a_3 x^3 + a_5 x^5 + ...$$

with two sets of coefficients computed from the recurrence formula as:

$$a_0; a_2 = \frac{-\lambda_2}{1 \cdot 2} a_0; a_4 = \frac{2 \cdot 3 - \lambda_4}{3 \cdot 4} a_2; ...$$

and

$$a_1; a_3 = \frac{1 \cdot 2 - \lambda_3}{2 \cdot 3} a_1; a_5 = \frac{3 \cdot 4 - \lambda_5}{4 \cdot 5} a_3; ...$$

and so on. To find the eigenvalues in general, we need to substitute the solution into the differential equation. Let us execute this for the $i = 1$ solution component

$$y_1(x) = a_1 x,$$

by ignoring the constant coefficient. Then

$$(1 - x^2)(x)'' - 2x(x)' + \lambda_1 x = (1 - x^2) \cdot 0 - 2x \cdot 1 + \lambda_1 x = -2x + 2x = 0,$$

which verifies the first eigenvalue that we obtained from the earlier equation. This may be written as

$$\lambda_1 = 2 = 1 \cdot (1 + 1).$$

This form also fits the trivial eigenvalue of λ_0 as $0 \cdot (0 + 1) = 0$. From that observation, we hypothesize that the eigenvalues are of the form

$$\lambda_i = i(i + 1), i = 0, 1, 2, ...,$$

Then $\lambda_2 = 2(2 + 1) = 6$ and the second solution component is

$$y_2 = a_0 \cdot 1 + \frac{-\lambda_2}{1 \cdot 2} a_0 \cdot x^2 = a_0(1 - 3x^2).$$

Ignoring the constant coefficient and substituting into the equation

$$(1 - x^2)(1 - 3x^2)'' - 2x(1 - 3x^2)' + \lambda_2(1 - 3x^2) = (1 - x^2)(-6) - 2x(-6x) + \lambda_2 x =$$

$$= -6 + 18x^2 + \lambda_2(1 - 3x^2) = 0.$$

which is true if $\lambda_2 = 6$ as expected. Similarly for the next solution component we obtain

$$a_3 = \frac{1 \cdot 2 - \lambda_3}{2 \cdot 3} a_1 = \frac{1 \cdot 2 - 3 \cdot 4}{2 \cdot 3} a_1 = -\frac{5}{3} a_1,$$

hence

$$y_3 = a_1 x + a_3 x^3 = a_1 \left(x - \frac{5}{3} x^3 \right).$$

The same process may be continued for more solution components, but it is not followed here further because a more efficient recursive formulation will be presented later.

Before doing so, we need to address the free coefficients. They may be resolved by enforcing certain boundary conditions at the ends of the interval, or applying the normalization constraint shown and used earlier. The Legendre polynomials' most-known form is obtained by enforcing the boundary conditions

$$y_i(1) = 1,$$

and

$$y_i(-1) = (-1)^i.$$

These result in the polynomials

$$Le_0 = 1,$$

$$Le_1 = x,$$

$$Le_2 = \frac{3}{2} x^2 - \frac{1}{2},$$

and

$$Le_3 = \frac{5}{2} x^3 - \frac{3}{2} x.$$

The first four Legendre polynomials are shown graphically in Figure 5.1 demonstrating the satisfaction of the boundary conditions.

It is easy to see that the odd Legendre polynomials are anti-symmetric

$$Le_k(x) = -Le_k(-x), k = 2i + 1,$$

and the even members are symmetric with respect to $x = 0$.

Higher order Legendre polynomials may be easily obtained from their recurrence relation

$$Le_{k+1}(x) = \frac{2k+1}{k+1} x Le_k(x) - \frac{k}{k+1} Le_{k-1}(x).$$

For illustration of the process of obtaining the higher order polynomials, the first six Legendre polynomials are summarized in Table 5.1.

A recurrence between the derivatives of Legendre functions is also useful

$$Le'_{k+1}(x) - Le'_{k-1}(x) = (2k+1)Le_k(x).$$

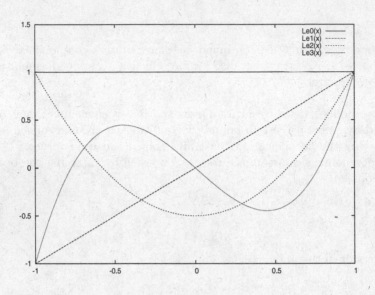

FIGURE 5.1 Legendre polynomials

TABLE 5.1
Legendre polynomials

i	$Le_i(x)$
0	1
1	x
2	$\frac{1}{2}(3x^2 - 1)$
3	$\frac{1}{2}(5x^3 - 3x)$
4	$\frac{1}{8}(35x^4 - 30x^2 + 3)$
5	$\frac{1}{8}(63x^5 - 70x^3 + 15x)$

Finally, as anticipated from the earlier part of the chapter, the Legendre polynomials are also orthogonal

$$\int_{-1}^{1} Le_i(x)Le_j(x)dx = 0; i \neq j,$$

and

$$\int_{-1}^{1} \left(Le_i(x)\right)^2 dx = \frac{2}{2i + 1}.$$

A very important application of the Legendre polynomials is the use of their zeroes as the locations of the so-called Gauss points. The points are the sampling locations of the numerical integration process of

$$\int f(x)dx = \sum_{k=1}^{n} w_{n,k} f(x_{n,k}),$$

also known as Gauss quadrature. The $x_{n,k}$ locations are the zeroes of the n-th Legendre polynomials:

$$Le_1(x) = 0 \rightarrow x_{1,1} = 0,$$

$$Le_2(x) = 0 \rightarrow x_{2,1} = -\frac{1}{\sqrt{3}}, x_{2,2} = -x_{2,1},$$

and

$$Le_3(x) = 0 \rightarrow x_{3,1} = \frac{3}{\sqrt{5}}, x_{3,2} = 0, x_{3,3} = -x_{3,1}.$$

The $w_{n,k}$ weights are computed by integrating Lagrange polynomials anchored by the zeroes of the Legendre polynomials over the interval of integration.

The importance of Sturm-Liouville problems lies in the fact that depending on the choice of the functions $p(x), q(x), r(x)$, a family of influential polynomials may be generated as the eigensolutions. The most known and used are, besides the Legendre polynomials above, the Bessel functions, Hermite, Chebyshev and Laguerre polynomials. They also exhibit orthogonality, symmetry and recursion properties of their own kind.

Besides the above mentioned, sometimes called classical problems, one can formulate specific problems for mathematical modeling of some physical phenomenon. For example, the Sturm-Liouville problem using $p(x)$ as the stiffness and $r(x)$ as the density of a flexible beam leads to the mathematical model of the longitudinal vibrations of the beam, a subject of Section 11.4.1.

Finally, Sturm-Liouville problems may also be presented with functions of multiple variables. For example, in three dimensions the equation becomes

$$-\nabla \left(p(x,y,z) \nabla u(x,y,z) \right) + q(x,y,z)u(x,y,z) = \lambda u(x,y,z)$$

leading to various elliptic partial differential equations that all have engineering implications. [8]

5.5 Exercises

Create the differential equations corresponding to the following Sturm-Liouville problems.

1. $p(x) = x, q(x) = -\frac{1}{x}, r(x) = x$.
Boundary conditions: $x_0 = 0, x_1 = \infty, y(x_0) = y(x_1) = 0$.

2. $p(x) = \sqrt{1 - x^2}, q(x) = 0, r(x) = \frac{1}{\sqrt{1-x^2}}$.
Boundary conditions: $x_0 = -1, x_1 = \infty, y(x_0) = y(x_1) = 0$.

3. $p(x) = e^{\frac{-x^2}{2}}, q(x) = 0, r(x) = e^{-\frac{x^2}{2}}$.
Boundary conditions: $x_0 = -\infty, x_1 = \infty, y(x_0) = y(x_1) = 0$.

4. $p(x) = xe^{-x}, q(x) = 0, r(x) = e^{-x}$.
Boundary conditions: $x_0 = 0, x_1 = \infty, y(x_0) = y(x_1) = 0$.

5. $p(x) = 1 - x^2, q(x) = 0, r(x) = 1$.
Boundary conditions: $y(0) = -1, y(1) = 1, y(x_0) = y(x_1) = 0$.

6

Analytic solutions

This chapter presents a handful of analytic methods for solving variational problems. They include the methods of Laplace transformation, d'Alembert's separation of variables techniques, the complete integrals and Poisson's integral formula. The method of gradients, with an illustrative example, concludes the chapter.

6.1   Laplace transform solution

The first method we discuss in this chapter transforms the original variational problem by applying Laplace transformation and producing an auxiliary differential equation [12].

Let us consider the variational problem of

$$I(t, x) = \int f(t, x)dt = \text{extremum},$$

and apply the Laplace transform to the function as

$$\int_0^\infty e^{-st} f(t, x)dt.$$

During this transform we regard t as the independent variable and x as a parameter. Note that the transformation of the boundary conditions is also required to obtain the complete auxiliary problem.

Let us illustrate this by the Euler-Lagrange differential equation of one spatial and one temporal independent variable of the form

$$a^2 \frac{\partial^2 u}{\partial x^2} - \frac{\partial u}{\partial t} = 0,$$

which is the one-dimensional heat equation, subject of deeper discussion in Section 11.6. We apply the initial condition

$$u(t = 0, x) = 0,$$

and boundary conditions

$$u(t, x = 0) = 0$$

and

$$u(t, x = 1) = B.$$

Executing the Laplace transform on the boundary conditions results in

$$\overline{u}(s, 0) = \int_0^\infty e^{-st} u(t, 0) dt = \int_0^\infty e^{-st} 0 \, dt = \frac{0}{s} = 0,$$

and

$$\overline{u}(s, 1) = \int_0^\infty e^{-st} u(t, 1) dt = \int_0^\infty e^{-st} B \, dt = \frac{B}{s}.$$

Transforming the yet unknown solution as

$$\overline{u}(s, x) = \int_0^\infty e^{-st} u(t, x) dt,$$

we produce the auxiliary equation in the form of

$$a^2 \frac{d^2 \overline{u}}{dx^2} - s\overline{u} = 0.$$

This equation is now ordinary; hence, it is easier to solve, demonstrating the advantage of this method. By integrating twice and applying Euler's formulae to eliminate complex terms, the solution of the auxiliary equation becomes

$$\overline{u}(s, x) = \frac{B \sinh(x \frac{\sqrt{s}}{a})}{s \sinh(\frac{\sqrt{s}}{a})}.$$

Finally, inverse Laplace transformation [2] yields the solution of the original problem in the form of

$$u(t, x) = B \left(x + \frac{2}{\pi} \sum_{k=1}^\infty \frac{(-1)^k}{k} e^{-(k\pi a)^2 t} \sin(k\pi x) \right).$$

This is the analytic solution of the one-dimensional heat conduction problem with constant temperature (B) at the boundary. The two- and three-dimensional heat conduction problems will be the subject of further discussion in the next section.

Let us now consider the variational problem of

$$I = \int_0^1 \left(\left(\frac{\partial u}{\partial t} \right)^2 - \left(\frac{\partial u}{\partial x} \right)^2 \right) dt = \text{extremum},$$

whose temporal derivative in the corresponding Euler-Lagrange equation is also of second order

$$a^2 \frac{\partial^2 u}{\partial x^2} - \frac{\partial^2 u}{\partial t^2} = 0.$$

The initial conditions are

$$u(t = 0, x) = 0, \frac{\partial u}{\partial t}(t = 0, x) = 0.$$

The boundary conditions are

$$u(t, x = 0) = 0, \frac{\partial u}{\partial x}(t, x = 1) = B.$$

The boundary conditions are transformed again as

$$\overline{u}(x = 0) = 0$$

and

$$\frac{d\overline{u}}{dx}(x = 1) = \frac{B}{s},$$

where s is the Laplace variable. The auxiliary equation becomes an ordinary differential equation of

$$\frac{d^2\overline{u}}{dx^2} = \frac{s^2}{a^2}\overline{u},$$

Integrating this equation, we obtain the result in the form

$$u(s, x) = \frac{B}{s^2}\frac{\sinh\left(\frac{sx}{a}\right)}{\cosh\left(\frac{s}{a}\right)}.$$

Finally, the inverse transformation yields the solution of the original problem at any point in the domain at any time as

$$u(t, x) = B\left(x - \frac{8}{\pi^2}\sum_{k=0}^{\infty}\frac{(-1)^k}{2k+1}\sin\left(\frac{\pi x}{2}(2k+1)\right)\cos\left(\frac{\pi a t}{2}(2k+1)\right)\right).$$

This is the analytic solution to the problem of the compression of a unit length beam along its longitudinal axis. The coefficient a and boundary condition B represent the physical characteristics of the beam and the problem, respectively. They will be introduced in connection with the solution of the axial vibration of a beam, presented in Section 11.4.1.

6.2   d'Alembert's solution

We now address the two-dimensional version of the problem of the last section

$$\frac{\partial u}{\partial t} = h^2\left(\frac{\partial^2 u}{\partial x^2} + \frac{\partial^2 u}{\partial y^2}\right).$$

We impose uniformly zero boundary conditions as

$$u(x, 0, t) = u(x, b, t) = 0; 0 \le x \le a,$$

and

$$u(0, y, t) = u(a, y, t) = 0; 0 \le y \le b.$$

The initial solution is given as a non-zero function of the spatial coordinates

$$u(x, y, 0) = f(x, y).$$

d'Alembert's separation of variables method seeks a solution in the form of

$$u(x, y, t) = e^{-\lambda t} u_1(x) u_2(y),$$

where λ is a yet unknown constant. Substitution and differentiation yield

$$-\lambda u_1 u_2 = h^2 (u_1'' u_2 + u_1 u_2'').$$

Conveniently reordering produces

$$\frac{u_1''}{u_1} + \frac{\lambda}{h^2} = -\frac{u_2''}{u_2} = k^2,$$

where k is a constant since the left-hand side is independent of y and the right-hand side is independent of x. Introducing

$$q^2 = \frac{\lambda}{h^2} - k^2,$$

we obtain a system of ordinary differential equations:

$$u_1'' + q^2 u_1 = 0,$$

and

$$u_2'' + k^2 u_2 = 0.$$

Their solutions are obtained as

$$u_1(x) = a_1 \sin(qx) + b_1 \cos(qx),$$

and

$$u_2(y) = a_2 \sin(ky) + b_2 \cos(ky).$$

The boundary conditions imply that $b_1 = b_2 = 0$, as well as

$$\sin(qa) = 0$$

and

$$\sin(kb) = 0.$$

Here a, b are the original spatial boundaries. Due to the periodic nature of the trigonometric functions

$$q = \frac{m\pi}{a}, m = 1, 2, ..$$

and

$$k = \frac{n\pi}{b}, n = 1, 2, ..$$

Substituting produces the unknown variable as

$$\lambda_{mn} = h^2 \left(\left(\frac{m\pi}{a} \right)^2 + \left(\frac{n\pi}{b} \right)^2 \right),$$

and the solution function of

$$u(x, y, t) = \sum_{m=1}^{\infty} \sum_{n=1}^{\infty} c_{mn} e^{-\lambda_{mn} l} \sin \frac{m\pi x}{a} \sin \frac{n\pi y}{b}.$$

The final unknown coefficient c_{mn} is obtained by the satisfaction of the initial condition:

$$f(x, y) = \sum_{m=1}^{\infty} \sum_{n=1}^{\infty} c_{mn} \sin \frac{m\pi x}{a} \sin \frac{n\pi y}{b},$$

from which the value of

$$c_{mn} = \frac{4}{ab} \int_0^b \int_0^a f(x, y) \sin \frac{m\pi x}{a} \sin \frac{n\pi y}{b} dx dy$$

emerges. It is easy to generalize this solution to the three-dimensional problem of

$$\frac{\partial u}{\partial t} = h^2 \left(\frac{\partial^2 u}{\partial x^2} + \frac{\partial^2 u}{\partial y^2} + \frac{\partial^2 u}{\partial z^2} \right).$$

We impose uniformly zero boundary conditions on three spatial dimensions as

$$u(x, y, 0, t) = u(x, y, c, t) = 0; 0 \leq x \leq a, 0 \leq y \leq b,$$
$$u(x, 0, z, t) = u(x, b, z, t) = 0; 0 \leq x \leq a, 0 \leq z \leq c,$$

and

$$u(0, y, z, t) = u(a, y, z, t) = 0; 0 \leq y \leq b, 0 \leq z \leq c.$$

The initial solution is given as a non-zero function of the three spatial coordinates,

$$u(x, y, z, 0) = f(x, y, z).$$

The solution with

$$\lambda_{mnr} = h^2 \left(\left(\frac{m\pi}{a} \right)^2 + \left(\frac{n\pi}{b} \right)^2 + \left(\frac{r\pi}{c} \right)^2 \right),$$

becomes

$$u(x, y, z, t) = \sum_{m=1}^{\infty} \sum_{n=1}^{\infty} \sum_{r=1}^{\infty} c_{mnr} e^{-\lambda_{mnr} t} \sin \frac{m\pi x}{a} \sin \frac{n\pi y}{b} \sin \frac{r\pi z}{c}.$$

The coefficient of the solution is also a straightforward generalization as

$$c_{mnr} = \frac{8}{abc} \int_0^c \int_0^b \int_0^a f(x, y, z) \sin \frac{m\pi x}{a} \sin \frac{n\pi y}{b} \sin \frac{r\pi z}{c} dx dy dz.$$

These last two solutions are the analytic solutions to the two- and three-dimensional heat conduction problem. The physical derivation of these problems is the subject of Section 11.6. The computational solution of the two-dimensional problem will be further addressed in Chapter 12.

Let us now solve the variational problem of two spatial variables again,

$$I = \int_0^b \int_0^a \left(\frac{\partial u}{\partial t} \right)^2 - h^2 \left(\left(\frac{\partial u}{\partial x} \right)^2 + \left(\frac{\partial u}{\partial y} \right)^2 \right) dx dy = \text{extremum},$$

but with a temporal variable whose second derivative is present in the Euler-Poisson equation:

$$h^2 \left(\frac{\partial^2 u}{\partial x^2} + \frac{\partial^2 u}{\partial y^2} \right) = \frac{\partial^2 u}{\partial t^2}.$$

We assume constant boundary conditions:

$$u(0, y, t) = u(a, y, t) = 0; 0 \le y \le b,$$

and

$$u(x, 0, t) = u(x, b, t) = 0; 0 \le x \le a.$$

The a, b are the dimensions of the domain. We seek the solution in the separated form of

$$u(x, y, t) = e^{i\lambda t} v(x, y).$$

Note the presence of the imaginary unit i in the exponent for later convenience. We introduce the constant

$$k^2 = \frac{\lambda^2}{h^2}.$$

Substitution yields the new differential equation

$$\frac{\partial^2 v}{\partial x^2} + \frac{\partial^2 v}{\partial y^2} + k^2 v = 0.$$

The new boundary conditions are

$$v(0, y) = v(a, y) = 0; 0 \le y \le b,$$

and

$$v(x,0) = v(x,b) = 0; 0 \leq x \leq a.$$

Furthermore, we separate the variables of this equation as

$$v(x,y) = u_1(x)u_2(y).$$

This leads to the system of equations

$$\frac{1}{u_1}\frac{d^2u_1}{dx^2} = -\frac{1}{u_2}\frac{d^2u_2}{dy^2} - k^2 = -m^2.$$

Here m is another yet unknown constant. The now familiar system of ordinary differential equations arises again

$$\frac{d^2u_1}{dx^2} + m^2u_1 = 0,$$

and

$$\frac{d^2u_2}{dy^2} + q^2u_2 = 0,$$

with $q^2 = k^2 - m^2$. The new boundary conditions are

$$u_1(0) = u_1(a) = 0,$$

and

$$u_2(0) = u_2(b) = 0.$$

Following the road paved earlier in this section, the first equation yields

$$u_1(x) = A_1 \sin(mx),$$

with $ma = n\pi, n = 1, 2, 3, \ldots$ and the second equation

$$u_2(y) = A_2 \sin(qy),$$

with $qb = r\pi, r = 1, 2, 3, \ldots$. Exploiting the relation

$$k^2 = m^2 + q^2 = \pi^2 \left(\frac{n^2}{a^2} + \frac{r^2}{b^2} \right)$$

we obtain the original parameter of the transformation

$$\lambda_{nr} = h\pi\sqrt{\frac{n^2}{a^2} + \frac{r^2}{b^2}}.$$

Finally by substituting and using Euler's formulae we obtain

$$u(x,y,t) = \sum_{n=1}^{\infty}\sum_{r=1}^{\infty} c_{nr} \cos(\lambda_{nr}t) \sin\frac{n\pi x}{a} \sin\frac{r\pi y}{b}.$$

The initial displacement represented by $f(x, y)$ aids in finding the final coefficient as

$$c_{nr} = \frac{4}{ab} \int_0^b \int_0^a f(x, y) \sin \frac{n\pi x}{a} \sin \frac{r\pi y}{b} dx dy.$$

This is the analytic solution of the problem of the vibrating membrane. A more general solution of this problem with variable boundary conditions is presented in Chapter 11.

Finally, let us consider an Euler-Lagrange equation of the first order with many independent variables in the implicit form of

$$F(x_1, x_2, ...x_n, u, \frac{\partial u}{\partial x_1}, \frac{\partial u}{\partial x_2}, ... \frac{\partial u}{\partial x_n}) = 0,$$

whose generic solution is

$$u(x_1, x_2, ...x_n; a_1, a_2, ..., a_n) = 0.$$

The solution to such a problem may be found by a repeated use of the separation of variables and the constant k. Let us first separate one variable as

$$u(x_1, x_2, ...x_n) = u_1(x_1) + u_2(x_2, x_3, ...x_n).$$

This corresponds to the following differential equation

$$F_1(x_1, u_1, \frac{\partial u_1}{\partial x_1}) = F_2(x_2, x_3...x_n, u_2, \frac{\partial u_2}{\partial x_2}, \frac{\partial u_2}{\partial x_3}, ... \frac{\partial u_2}{\partial x_n}) = k^2.$$

The equation may be satisfied by solving a pair of equations with an unknown constant:

$$F_1(x_1, u_1, \frac{\partial u_1}{\partial x_1}) = k^2,$$

and

$$F_2(x_2, x_3...x_n, u_2, \frac{\partial u_2}{\partial x_2}, \frac{\partial u_2}{\partial x_3}, ... \frac{\partial u_2}{\partial x_n}) = k^2.$$

The first equation becomes an ordinary differential equation whose solution is easily obtained. The second equation may be further separated and the same process continued.

6.3   Complete integral solutions

For certain problems a complete integral solution is available [15]. The complete integral form presents a parametric family of general solutions. The particular solution of a specific problem can then be obtained from the general complete integral solution by selection of the parameters.

We will first demonstrate generating a complete integral solution by exploiting the concept of separation of variables introduced in the last section. For the simplicity of the discussion, and without loss of generality, we will do this with an example of only two independent variables. For a given

$$F(x, y, u, \frac{\partial u}{\partial x}, \frac{\partial u}{\partial y}) = 0,$$

the complete integral solution is of the form

$$u(x, y, a, b),$$

where the a, b are yet unknown coefficients. Let us generate the complete integral solution for the equation of

$$\left(\frac{\partial u}{\partial x}\right)^2 + \left(\frac{\partial u}{\partial y}\right)^2 = 1.$$

We seek a separated solution of the form

$$u(x, y) = u_1(x) + u_2(y).$$

The first differential equation with a constant k is then

$$F_1\left(x, u_1, \frac{du_1}{dx}\right) = \left(\frac{du_1}{dx}\right)^2 = k^2.$$

The solution comes by

$$du_1 = kdx,$$

from which

$$u_1 = kx + k_1$$

emerges. Similarly, the second, in this case also ordinary equation is

$$F_2\left(y, u_2, \frac{du_2}{dy}\right) = 1 - \left(\frac{du_2}{dy}\right)^2 = k^2.$$

The solution of

$$du_2 = \sqrt{1 - k^2}dy,$$

yields

$$u_2 = \sqrt{1 - k^2}y + k_2.$$

Finally, the complete integral solution for this problem is

$$u(x, y, a, b) = ax + \sqrt{1 - a^2}y + b,$$

with $a = k$ and $b = k_1 + k_2$. The complete integral solution satisfies the original problem

$$\left(\frac{\partial u}{\partial x}\right)^2 + \left(\frac{\partial u}{\partial y}\right)^2 = 1,$$

since

$$\frac{\partial u}{\partial x} = a, \frac{\partial u}{\partial y} = \sqrt{1 - a^2},$$

and

$$a^2 + 1 - a^2 = 1.$$

This is a two-parameter family of solutions from which any particular solution may be obtained. Surely any selection of the parameter b will satisfy the original equation. As far as the parameter a is concerned, selecting for example $a = 1/2$ will result in

$$\frac{\partial u}{\partial x} = \frac{1}{2}, \frac{\partial u}{\partial y} = \sqrt{1 - \frac{1}{4}},$$

and

$$\frac{1}{4} + 1 - \frac{1}{4} = 1.$$

When generating a complete integral solution, the separation strategy depends on the given differential equation. When second derivatives are also present, a product type separation may be used. For example, for the equation

$$\frac{\partial^2 u}{\partial x^2} - \frac{\partial u}{\partial y} = 0$$

the separated solution of the form

$$u(x, y) = u_1(x) \cdot u_2(y)$$

is recommended. The pair of differential equations in this scenario are

$$\frac{1}{u_1} \frac{d^2 u_1}{dx^2} = k^2,$$

and

$$\frac{1}{u_2} \frac{d u_2}{dy} = k^2.$$

The solution of this system is of the form

$$u(x, y) = (k_1 e^{kx} + k_2 e^{-kx}) e^{k^2 y}.$$

Note that three parameters are needed because of the presence of the second derivative. Since this is the complete integral solution, we have the freedom of choice of the parameters. By setting them all to unity, a particular solution emerges as

$$u(x, y) = (e^x + e^{-x})e^y = e^{x+y} + e^{y-x}.$$

To validate the solution, we compute

$$\frac{\partial^2 u}{\partial x^2} = e^{x+y} + e^{y-x}$$

and

$$\frac{\partial u}{\partial y} = e^{x+y} + e^{y-x},$$

whose difference is the desired zero.

Let us now consider simply using pre-computed complete integrals. Certain complete integral solutions actually contain integrals. Consider the non-homogeneous differential equation type with non-constant coefficients

$$a(x)\left(\frac{\partial u}{\partial x}\right)^2 + b(x)\left(\frac{\partial u}{\partial y}\right)^2 = f(x) + g(y).$$

Such problems have a complete integral solution of

$$u(x, y) = \int_0^x \sqrt{\frac{f(t) + a_1}{a(t)}}dt + \int_0^y \sqrt{\frac{y(t) - a_1}{b(t)}}dt + a_2.$$

For example, the equation

$$\left(\frac{\partial u}{\partial x}\right)^2 + \left(\frac{\partial u}{\partial y}\right)^2 = x + y$$

has a complete integral solution of the form

$$u(x, y) = \int_0^x \sqrt{t + a_1}dt + \int_0^y \sqrt{t - a_1}dt + a_2.$$

There are also rather specific, but practical problems where the partial derivatives occur in an exponential expression. The generic form of such problems is

$$\frac{\partial u}{\partial x} = f(x)\frac{\partial u}{\partial y} + g(x)e^{\frac{\partial u}{\partial y}}.$$

The complete integral solution of this problem is in the following form

$$u = a_1 \int_0^x f(t)dt + e^{a_1} \int_0^x g(t)dt + a_1 y + a_2.$$

For an example of this case, the equation

$$\frac{\partial u}{\partial x} = x^2\frac{\partial u}{\partial y} + xe^{\frac{\partial u}{\partial y}},$$

has a complete integral solution in the following form

$$u = a_1 x^3/3 + e^{a_1} x^2/2 + a_1 y + a_2.$$

Finding a particular solution from a complete integral solution is not always trivial. From the complete integral solution of the form

$$u = f(x_1, x_2, ..., x_n, a_1, a_2, ..., a_n)$$

the introduction of another set of coefficients as

$$\frac{\partial f}{\partial a_i} = b_i$$

results in a new complete integral solution of

$$u = f(a_1, a_2, ..., a_n, b_1, b_2, ..., b_n).$$

This may provide an easier way toward the particular solution form.

6.4   Poisson's integral formula

We consider Laplace's equation in two dimensions, which plays a fundamental role in mathematical physics:

$$\frac{\partial^2 u}{\partial x^2} + \frac{\partial^2 u}{\partial y^2} = 0.$$

We will assume a circular domain and use the polar coordinate form as

$$\frac{\partial^2 u}{\partial r^2} + \frac{1}{r}\frac{\partial u}{\partial r} + \frac{1}{r^2}\frac{\partial^2 u}{\partial \phi^2} = 0,$$

where $r = \sqrt{x^2 + y^2}, \phi = \arctan \frac{y}{x}$. Here the r is the radius and the ϕ is the polar angle. Using the separation of variables again, we seek the solution in the form of

$$u(r, \phi) = u_1(r)u_2(\phi),$$

with the notational convention also followed. Substituting into the equation, we obtain

$$\frac{1}{u_1(r)}(r^2\frac{d^2 u_1}{dr^2} + r\frac{du_1}{dr}) = -\frac{1}{u_2}\frac{d^2 u_2}{d\phi^2} = k^2,$$

where k^2 is the yet unknown coefficient. The resulting pair of ordinary differential equations becomes

$$r^2 \frac{d^2 u_1}{dr^2} + r \frac{du_1}{dr} - k^2 u_1 = 0$$

and

$$\frac{d^2 u_2}{d\phi^2} + k^2 u_2 = 0.$$

The general solutions of these equations were derived in an earlier section. For $k = 0$ the separate solutions are

$$u_{2,0}(\phi) = a_0 \phi + b_0$$

and

$$u_{1,0}(r) = c_0 \ln(r) + d_0.$$

The complete solution for the $k = 0$ case is

$$u_0(r, \phi) = (a_0 \phi + b_0)(c_0 \ln(r) + d_0).$$

In the case of $k \neq 0$, the separated solutions are

$$u_{2,k}(\phi) = a_k \cos(k\phi) + b_k \sin(k\phi),$$

and

$$u_{1,k}(r) = c_k r^k + d_k r^{-k}.$$

The solution of the problem is then

$$u_k(r, \phi) = (a_k \cos(k\phi) + b_k \sin(k\phi)) \left(c_k r^k + d_k r^{-k} \right); k \neq 0.$$

We assume a uniquely defined solution function; therefore,

$$u_k(r, \phi) = u_k(r, \phi + 2\pi); k \neq 0,$$

which implies that k can only be an integer. Executing the multiplication and introducing the products

$$\overline{a}_k = a_k c_k,$$

$$\overline{b}_k = b_k c_k,$$

$$\overline{c}_k = a_k d_k$$

and

$$\overline{d}_k = b_k d_k,$$

we obtain for $k \neq 0$

$$u_k(r, \phi) = \sum_{k=1}^{\infty} r^k \left(\overline{a}_k \cos(k\phi) + \overline{b}_k \sin(k\phi) \right) + \sum_{k=1}^{\infty} \frac{1}{r^k} \left(\overline{c}_k \cos(k\phi) + \overline{d}_k \sin(k\phi) \right).$$

The constants may be found by the boundary conditions. Dictating that the solution be non-zero and bounded at the origin implies that

$$a_0, c_0, \overline{c}_k, \overline{d}_k = 0.$$

Because the Laplace equation is linear and homogeneous, the solution is the sum of the $k = 0$ and $k \neq 0$ solutions:

$$u(r, \phi) = \overline{a}_0 + \sum_{k=1}^{\infty} r^k \left(\overline{a}_k \cos(k\phi) + \overline{b}_k \sin(k\phi) \right).$$

Here $\overline{a}_0 = b_0 d_0$. Let us impose another, external boundary condition at the outermost radius of our interest as

$$u(r_{max}, \phi) = f(\phi).$$

Substituting into the solution form, we get

$$f(\phi) = \overline{a}_0 + \sum_{k=1}^{\infty} r_{max}^k \left(\overline{a}_k \cos(k\phi) + \overline{b}_k \sin(k\phi) \right).$$

Hence the coefficients become

$$\overline{a}_0 = \frac{1}{2\pi} \int_0^{2\pi} f(\phi) d\phi,$$

$$\overline{a}_k = \frac{1}{r_{max}^k \pi} \int_0^{2\pi} f(\phi) \cos(k\phi) d\phi,$$

and

$$\overline{b}_k = \frac{1}{r_{max}^k \pi} \int_0^{2\pi} f(\phi) \sin(k\phi) d\phi.$$

Bringing the now resolved coefficients into the generic solution form and introducing a new integral variable ψ, we obtain

$$u(r, \phi) = \frac{1}{2\pi} \int_0^{2\pi} f(\psi) d\psi + \frac{1}{\pi} \sum_{k=1}^{\infty} \left(\frac{r}{r_{max}} \right)^k \left(\sin(k\phi) \int_0^{2\pi} f(\psi) \sin(k\psi) d\psi + \right.$$

$$\left. \cos(k\phi) \int_0^{2\pi} f(\psi) \cos(k\psi) d\psi \right).$$

Employing the algebraic identity of

$$\cos(\psi - \phi) = \cos(\phi) \cos(\psi) + \sin(\phi) \sin(\psi),$$

we can write

$$u(r, \phi) = \frac{1}{2\pi} \int_0^{2\pi} f(\psi) d\psi + \frac{1}{\pi} \sum_{k=1}^{\infty} \int_0^{2\pi} f(\psi) \left(\frac{r}{r_{max}} \right)^k \cos\left(k(\psi - \phi) \right) d\psi.$$

Since for $0 \leq r \leq r_{max}$ the series of

$$-\sum_{k=1}^{\infty} \left(\frac{r}{r_{max}}\right)^k \cos\left(k(\psi - \phi)\right)$$

is monotonically convergent, the order of the integration and summation may be changed. This results in the form:

$$u(r, \phi) = \frac{1}{\pi} \int_0^{2\pi} f(\psi) \left(\frac{1}{2} + \sum_{k=1}^{\infty} \left(\frac{r}{r_{max}}\right)^k \cos\left(k(\psi - \phi)\right)\right) d\psi.$$

Finally, we use Euler's formula to replace the cos term as

$$\cos\left(k(\psi - \phi)\right) = \frac{e^{ik(\psi-\phi)} + e^{-ik(\psi-\phi)}}{2}.$$

In the above expression, $i = \sqrt{-1}$ is the imaginary unit. Substituting the above and after some algebraic manipulations, we obtain

$$u(r, \phi) = \frac{1}{2\pi} \int_0^{2\pi} f(\psi) \frac{r_{max}^2 - r^2}{r_{max}^2 - 2r_{max}r\cos(\phi - \psi) + r^2} d\psi.$$

This formula is known as Poisson's integral formula. With this, the solution value of Laplace's equation on a bounded circular domain may be obtained at any radius $0 \leq r \leq r_{max}$ and at any angle $\phi + j2\pi, j = 0, 1, 2, ...,$ for a given boundary value function $f(\phi)$.

Laplace's equation also occurs in three-dimensional form as

$$\frac{\partial^2 u}{\partial x^2} + \frac{\partial^2 u}{\partial y^2} + \frac{\partial^2 u}{\partial z^2} = 0.$$

Assuming a circular domain as above, the z dimension's presence leads us to use cylindrical coordinates

$$\frac{\partial^2 u}{\partial r^2} + \frac{1}{r}\frac{\partial u}{\partial r} + \frac{1}{r^2}\frac{\partial^2 u}{\partial \phi^2} + \frac{\partial^2 u}{\partial z^2} = 0.$$

Following the method laid out above, we seek the solution in the form of

$$u(r, \phi) = u_1(r)u_2(\phi)u_3(z).$$

Substitution yields

$$\frac{1}{u_3}\frac{d^2 u_3}{dz^2} = -\frac{1}{u_1}\frac{d^2 u_1}{dr^2} - \frac{1}{u_1 r}\frac{du_1}{dr} - \frac{1}{u_2 r^2}\frac{d^2 u_2}{d\phi^2}.$$

Relying on the insight gained in the last sections using an unknown coefficient, for the left-hand side we choose the solution of

$$\frac{1}{u_3}\frac{d^2 u_3}{dz^2} = k^2.$$

This choice yields

$$u_3(z) = c_1 e^{kz} + c_2 e^{-kz}.$$

The right-hand side of the problem may be written as

$$\frac{r^2}{u_1} \frac{d^2 u_1}{dr^2} + \frac{r}{u_1} \frac{du_1}{dr} + k^2 r^2 = -\frac{1}{u_2} \frac{d^2 u_2}{d\phi^2} = m^2,$$

where m is another yet unknown constant. In order to have a solution as a uniquely defined function in ϕ, m is again an integer. The right-hand equation becomes an ordinary differential equation as

$$\frac{d^2 u_2}{d\phi^2} + m^2 u_2 = 0,$$

whose solution is

$$u_2(\phi) = c_3 \cos(m\phi) + c_4 \sin(m\phi).$$

Finally the remaining equation is of the Bessel kind:

$$r^2 \frac{d^2 u_1}{dr^2} + r \frac{du_1}{dr} + (k^2 r^2 - m^2) u_1 = 0.$$

The solution of such differential equation when m is not an integer is of the form

$$u_1(r) = c_5 J_m(kr) + c_6 J_{-m}(kr),$$

where J are the Bessel functions of the first kind, defined by the formula

$$J_m(x) = \sum_{n=0}^{\infty} \frac{(-1)^n}{n!(n+m)!} \left(\frac{x}{2}\right)^{m+2n}.$$

This is a convergent series for any $x = kr$ value. The J_{-m} function in the expression is simply defined by

$$J_{-m}(x) = (-1)^m J_m(x).$$

However, in our case, m is an integer and so the solution is

$$u_1(r) = c_5 J_m(kr) + c_6 Y_m(kr).$$

The Bessel function of the second kind is defined as

$$Y_m(x) = \lim_{p \to m} \frac{\cos(p\pi) J_p(x) - J_{-p}(x)}{\sin(p\pi)}.$$

The limit is needed since the denominator is zero for any integer multiple of π. Therefore, this function is infinite at the origin; hence, to assure that at $r = 0$ we have a bounded solution, we choose $c_6 = 0$. Then the term with Y_m

drops out and the solution of this equation becomes

$$u_1(r) = c_5 J_m(kr).$$

When m is an integer, Bessel functions of the first kind can also be computed from integral formulae:

$$J_m(x) = \frac{1}{\pi} \int_0^\pi \cos\left(x\sin(t)\right)\cos(mt)dt$$

when m is even, and

$$J_m(x) = \frac{1}{\pi} \int_0^\pi \sin\left(x\sin(t)\right)\sin\left((m+1)t\right)dt$$

when m is odd.

Finally, rejoining the separated solutions, we obtain

$$u(r,\phi,z) = \sum_{m=0}^\infty \Big(e^{kz}\left(a_m\cos(m\phi) + b_m\sin(m\phi)\right) +$$

$$+ e^{-kz}\left(d_m\cos(m\phi) + e_m\sin(m\phi)\right)\Big)J_m(kr).$$

Here a_m, b_m, d_m, e_m are various products of the above c_k constants to be specified by the boundary conditions. This is the general solution of the three-dimensional Laplace equation in cylindrical coordinates.

6.5   Method of gradients

The final method in this class of solutions is that of the gradients. Let us focus on the first order variational problem

$$I(y) = \int_0^1 f(x, y, y')dx = \text{extremum}.$$

Denoting W as the space of absolutely continuous functions, introduce the linear subspace

$$\overline{W} = \Big(y(x)\Big|y \in W, y(0) = 0\Big)$$

on which the norm is defined by the scalar product

$$(y_1, y_2) = y_1(0)y_2(0) + \int_0^1 y_1'(x)y_2'(x)dx.$$

Borrowing from calculus, we will define the derivative of the functional as

$$I'(y) = \lim_{t \to 0} \frac{I(y + t\eta) - I(y)}{t},$$

and this limit exists for all variations $\eta \in \overline{W}$. For $I'(y) \in \overline{W}$ it follows that the derivative, also called the functional gradient, is of the form [7]

$$I'(y) = \int_0^x \left(\int_t^1 \frac{\partial}{\partial y} f\left(s, y(s), y'(s)\right) ds + \frac{\partial}{\partial y'} f\left(t, y(t), y'(t)\right) \right) dt.$$

With this gradient, we construct an iteration procedure as follows. Starting from an initial solution $y_1(x)$, compute successive approximations $y_i(x), i = 1, 2, \ldots$. In every iteration, the gradient is computed at the current solution:

$$I'(y_i) = \int_0^x \left(\int_t^1 \frac{\partial}{\partial y_i} f\left(s, y_i(s), y_i'(s)\right) ds + \frac{\partial}{\partial y_i'} f\left(t, y_i(t), y_i'(t)\right) \right) dt.$$

It is followed by finding the distance scale α_i as solution of

$$min_{\alpha \geq 0} I\left(y_i - \alpha I'(y_i)\right).$$

With this, the next iteration is computed as a scaled step into the direction of the gradient as

$$y_{i+1} = y_i - \alpha_i I'(y_i).$$

The sequence of

$$I(y_1) \geq I(y_2) \geq \ldots \geq I(y_i)$$

is continued until one of two conditions is satisfied. If $I(y_i)$ becomes zero, then $y_i(x)$ is an extremal and satisfies the Euler-Lagrange differential equation of the problem. Otherwise, the process terminates when it no longer progresses, measured by

$$\frac{I(y_i) - I(y_{i+1})}{I(y_i)} < \epsilon,$$

where ϵ is an appropriately chosen small number. Most of the time this terminates the procedure and the exact extremum and solution is not reached.

We illustrate the method by computing the solution of the variational problem

$$I(y) = \int_0^1 (2xy + y^2 + y'^2) dx = \text{extremum},$$

with initial condition $y(0) = 0$, and start the procedure from $y_1(x) = 0$. The first step is posed as

$$I'(y_1) = \int_0^x \left(\int_t^1 \frac{\partial}{\partial y} f\left(s, y_1(s), y_1'(s)\right) ds + \frac{\partial}{\partial y'} f\left(t, y_1(t), y_1'(t)\right) \right) dt.$$

This results in

$$I'(y_1) = \int_0^x \left(\int_t^1 2s + 2y_1(s)ds + 2y_1'(t) \right) dt.$$

The first term of the double integral is

$$\int_0^x \int_t^1 2sdsdt = \int_0^x s^2 \Big|_t^1 dt = \int_0^x (1 - t^2)dt = x - \frac{x^3}{3}.$$

The second term is zero due to the starting solution being zero. Similarly the single integral

$$\int_0^x \frac{\partial f}{\partial y_1'}dt = \int_0^x 2y_1'(t)dt = 2y_1(x) = 0,$$

is also zero. Hence the gradient at this stage is

$$I'(y_1) = x - \frac{x^3}{3}.$$

To find α_1 as the solution of

$$min_{\alpha \geq 0} I(y_1 - \alpha I'(y_1)),$$

compute

$$I(y_1 - \alpha I'(y_1)) = I\left(-\alpha(x - \frac{x^3}{3}) \right) =$$

$$= \int_0^1 \left(-2x\alpha(x - \frac{x^3}{3}) + \alpha^2(x - \frac{x^3}{3})^2 + \alpha^2(1 - x^2)^2 \right) dx.$$

Executing the posted integrations and substitutions yields

$$\frac{4}{15}(\frac{59}{21}\alpha^2 - \alpha),$$

whose minimum produces

$$\alpha_1 = \frac{21}{118}.$$

From this the next iterative solution becomes

$$y_2(x) = -\frac{21}{118}(x - \frac{x^3}{3}),$$

and the next iteration of the extremum is

$$I(y_2) = -\frac{21}{295}.$$

Continuing with

$$I'(y_2) = \int_0^x \left(\int_t^1 \frac{\partial}{\partial y} f(s, y_2(s), y_2'(s)) \, ds + \frac{\partial}{\partial y'} f(t, y_2(t), y_2'(t)) \right) dt$$

would produce the next solution iteration in the form of

$$y_3(x) = y_2(x) - \alpha_2 I'(y_2).$$

The process continues until the process is terminated by either of the detection mechanisms described above.

The gradient method is the basis for optimization solutions of application problems in many industries, for example in structural engineering. In the latter case, however, the gradient is computed by evaluating the solution function at discrete locations in the solution domain and applied in connection with finite element discretization (to be discussed in detail in Chapter 12).

6.6   Exercises

Use d'Alembert's solution to solve Example 1.

1. $\left(\frac{\partial u}{\partial t}\right)^2 = (\frac{\partial u}{\partial x})^2 + (\frac{\partial u}{\partial y})^2.$

Use the complete integration approach to solve Problems 2 and 3.

2. $\left(\frac{\partial u}{\partial x}\right)^2 + (\frac{\partial u}{\partial y})^2 = 2.$

3. $\left(\frac{\partial u}{\partial x}\right)^2 + (\frac{\partial u}{\partial y})^2 = x + y.$

Execute two iterations of the gradient method on Problems 4 and 5.

4. $\int_0^1 (y'^2 + y^2)dx = \text{extremum}; y_1(x) = 1.$

5. $\int_0^1 (y'^2 - y^2 - 2xy)dx = \text{extremum}; y_1(x) = 0.$

7

Approximate methods

This chapter addresses problems that may not be easily solved by analytic techniques, if solvable at all. Hence, before we embark on applications in later chapters, we discuss techniques that provide approximate solutions for such problems.

The discussion starts with the classical method of this class, the Euler method, and the most influential method, that of Ritz's. The methods of Galerkin and Kantorovich follow, both described in [10]. They could be considered extensions of Ritz's. Finally, the boundary integral and the finite element methods, the most well-known by engineers and used in the industry, conclude the chapter.

7.1 Euler's method

Euler proposed a numerical solution for the variational problem of

$$I(y) = \int_{x_0}^{x_n} f(x, y, y')dx = \text{extremum}$$

with the boundary conditions

$$y(x_0) = y_0; y(x_n) = y_n,$$

by subdividing the interval of the independent variable as

$$x_i = x_0 + i\frac{x_n - x_0}{n}; i = 1, 2, \ldots, n.$$

Introducing

$$h = \frac{x_n - x_0}{n},$$

the functional may be approximated as

$$I(y_i) = \int_{x_0}^{x_1} f(x_i, y_i, y_i') = h\sum_{i=1}^{n-1} f(x_0 + ih, y_i, \frac{y_{i+1} - y_i}{h})dx = \text{extremum}.$$

Here the approximated solution values y_i are the unknowns and the extremum may be found by differentiation:

$$\frac{\partial I}{\partial y_i} = 0.$$

The process is rather simple and follows from Euler's other work in the numerical solution of differential equations. For illustration, we consider the following problem:

$$I(y) = \int_0^1 (2xy + y^2 + y'^2)dx = \text{extremum},$$

with the boundary conditions

$$y(0) = y(1) = 0.$$

Let us subdivide the interval into $n = 5$ equidistant segments with

$$h = 0.2,$$

and

$$x_i = 0.2i.$$

The approximate functional with the appropriate substitutions becomes

$$I(y_i) = 0.2 \sum_{i=1}^4 (0.4iy_i + y_i^2 + (5\,(y_{i+1} - y_i))^2).$$

The computed partial derivatives are

$$\frac{\partial I}{\partial y_1} = 0.2(0.4 + 2y_1 - \frac{2(y_2 - y_1)}{0.04}) = 0,$$

$$\frac{\partial I}{\partial y_2} = 0.2(0.8 + 2y_2 - \frac{2(y_3 - y_2)}{0.04} + \frac{2(y_2 - y_1)}{0.04}) = 0,$$

$$\frac{\partial I}{\partial y_3} = 0.2(1.2 + 2y_3 - \frac{2(y_4 - y_3)}{0.04} + \frac{2(y_3 - y_2)}{0.04}) = 0,$$

and

$$\frac{\partial I}{\partial y_4} = 0.2(1.6 + 2y_4 + \frac{2y_4}{0.04} + \frac{2(y_4 - y_3)}{0.04}) = 0.$$

This system of four equations yields the values of the approximate solution. The analytic solution of this problem is

$$y(x) = -x + e\frac{e^x - e^{-x}}{e^2 - 1}.$$

The comparison of the Euler solution (y_i) and the analytic solution $(y(x_i))$ at the four discrete points is shown in Table 7.1.

TABLE 7.1
Accuracy of Euler's
method

i	x_i	y_i	$y(x_i)$
1	0.2	-0.0286	-0.0287
2	0.4	-0.0503	-0.0505
3	0.6	-0.0580	-0.0583
4	0.8	-0.0442	-0.0444

The boundary solutions of $y(0)$ and $y(1) = 0$ are not shown since they are in full agreement by definition.

7.2   Ritz's method

Let us consider the variational problem of

$$I(y) = \int_{x_0}^{x_1} f(x, y, y')dx = \text{extremum},$$

under the boundary conditions

$$y(x_0) = y_0; y(x_1) = y_1.$$

The Ritz method is based on an approximation of the unknown solution function with a linear combination of certain basis functions. Finite element or spline-based approximations are the most commonly used and will be subject of detailed discussion in Chapters 9 and 11. Let the unknown function be approximated with

$$\overline{y}(x) = \alpha_0 b_0(x) + \alpha_1 b_1(x) + \ldots + \alpha_n b_n(x),$$

where the basis functions are also required to satisfy the boundary conditions and the coefficients are yet unknown. Substituting the approximate solution into the variational problem results in

$$I(\overline{y}) = \int_{x_0}^{x_1} f(x, \overline{y}, \overline{y}')dx = \text{extremum}.$$

In order to reach an extremum of the functional, it is necessary that the derivatives with respect to the unknown coefficients vanish:

$$\frac{\partial I(\overline{y})}{\partial \alpha_i} = 0; i = 0, 1, \ldots, n.$$

It is not intuitively clear that the approximated function approaches the extremum of the original variational problem, but it has been proven, for example in [10].

Let us just illustrate the process with a small analytic example. Consider the variational problem of

$$I(y) = \int_0^1 y'^2(x)dx = \text{extremum},$$

with the boundary conditions

$$y(0) = y(1) = 0,$$

and constraint of

$$\int_0^1 y^2(x)dx = 1.$$

Since this is a constrained problem, we apply the Lagrange multiplier technique and rewrite the variational problem as

$$I(y) = \int_0^1 (y'^2(x) - \lambda y^2)dx = \text{extremum}.$$

Let us use, for example, the basis functions of

$$b_0(x) = x(x-1)$$

and

$$b_1(x) = x^2(x-1).$$

It is trivial to verify that these also obey the boundary conditions. The approximated solution function is

$$\bar{y} = \alpha_0 x(x-1) + \alpha_1 x^2(x-1).$$

The functional of the constrained, approximated variational problem is

$$I(\bar{y}) = \int_0^1 (\bar{y}'^2 - \lambda \bar{y}^2)dx.$$

Evaluating the integral yields

$$I(\bar{y}) = \frac{1}{3}(\alpha_0^2 + \alpha_0\alpha_1 + \frac{2}{5}\alpha_1^2) - \lambda(\frac{1}{30}\alpha_0^2 + \frac{1}{30}\alpha_0\alpha_1 + \frac{1}{105}\alpha_1^2).$$

The extremum requires the satisfaction of

$$\frac{\partial I}{\partial \alpha_0} = \alpha_0\left(\frac{2}{3} - \frac{\lambda}{15}\right) + \alpha_1\left(\frac{1}{3} - \frac{\lambda}{30}\right) = 0$$

and

$$\frac{\partial I}{\partial \alpha_1} = \alpha_0 \left(\frac{1}{3} - \frac{\lambda}{30} \right) + \alpha_1 \left(\frac{4}{15} - \frac{2\lambda}{105} \right) = 0.$$

A non-trivial solution of this system of equations is obtained by setting its determinant to zero, resulting in the following quadratic equation

$$\frac{\lambda^2 - 52\lambda + 420}{6300} = 0.$$

Its solutions are

$$\lambda_1 = 10; \lambda_2 = 42.$$

Using the first value and substituting into the second condition yield

$$\alpha_1 = 0$$

with arbitrary α_0. Hence

$$\overline{y}(x) = \alpha_0 x(x - 1).$$

The condition

$$\int_0^1 \overline{y}^2 dx = \int_0^1 \alpha_0^2 x^2 (x - 1)^2 dx = 1$$

results in

$$\alpha_0 = \pm\sqrt{30}.$$

The approximate solution of the variational problem is

$$\overline{y}(x) = \pm\sqrt{30} x(x - 1).$$

It is very important to point out that the solution obtained as a function of the chosen basis functions is not the analytic solution of the variational problem. For this particular example, the corresponding Euler-Lagrange differential equation is

$$y'' + \lambda y = 0$$

whose analytic solution, based on Section 5.3, is

$$y = \pm\sqrt{2} \sin(\pi x).$$

Figure 7.1 compares the analytic and the approximate solutions and plots the error of the latter.

The figure demonstrates that the Ritz solution satisfies the boundary conditions and shows acceptable differences in the interior of the interval. Finally,

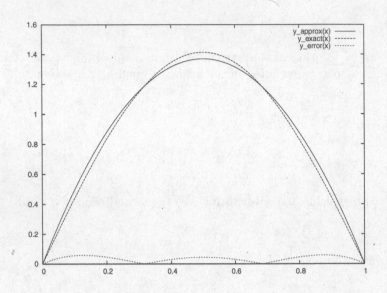

FIGURE 7.1 Accuracy of the Ritz solution

the variational problem's extremum is computed for both cases. The analytical solution is based on the derivative

$$y' = \sqrt{2}\pi \cos(\pi x),$$

and obtained as

$$\int_0^1 y'^2(x)dx = 2\pi^2 \int_0^1 \cos^2(\pi x)dx = \pi^2 = 9.87.$$

The Ritz solution's derivative is

$$\overline{y}' = -\sqrt{30}(2x - 1),$$

and the approximate extremum is

$$\int_0^1 \overline{y}'^2(x)dx = 30 \int_0^1 (2x - 1)^2 dx = \frac{30}{3} = 10.$$

The approximate extremum is slightly higher than the analytic extremum, but by only a very acceptable error.

7.3 Galerkin's method

The difference between Ritz's method and that of Galerkin's is in the fact that the latter addresses the differential equation form of a variational problem. Galerkin's method minimizes the residual of the differential equation integrated over the domain with a weight function; hence, it is also called the method of weighted residuals.

This difference lends more generality and computational convenience to Galerkin's method. Let us consider a linear differential equation in two variables:

$$L\left(u(x,y)\right) = 0$$

and apply Dirichlet boundary conditions. Galerkin's method is also based on the Ritz approximation of the solution as

$$\overline{u} = \sum_{i=1}^{n} \alpha_i b_i(x,y),$$

in which case, of course there is a residual of the differential equation

$$L(\overline{u}) \neq 0.$$

Galerkin proposed using the basis functions of the approximate solution also as the weights, and required the integral to vanish with a proper selection of the coefficients:

$$\int\int_D L(\overline{u}) b_j(x,y) dx dy = 0; j = 1, 2, \ldots, n.$$

This yields a system for the solution of the coefficients as

$$\int\int_D L\left(\sum_{i=1}^{n} \alpha_i b_i(x,y)\right) b_j(x,y) dx dy = 0; j = 1, 2, \ldots, n.$$

This is also a linear system and produces the unknown coefficients α_i.

For illustration of this method, let us consider the example problem already solved in Section 1.1 via its Euler-Lagrange differential equation:

$$I = \int_0^1 (\frac{1}{2}y'^2 + (x+1)y)dx = \text{extremum},$$

with boundary conditions

$$y(0) = 0, y(1) = 1.$$

We seek the approximate solution using power basis functions up to the third order in the form of

$$\overline{y}(x) = c_1 + c_2 x + c_3 x^2 + c_4 x^3.$$

Immediately applying the boundary conditions enables us to simplify the approximate solution

$$\overline{y}(0) = 0 \rightarrow c_1 = 0$$

and

$$\overline{y}(1) = 1 \rightarrow c_2 + c_3 + c_4 = 1$$

results in

$$c_2 = 1 - c_3 - c_4,$$

and the approximate solution becomes

$$\overline{y}(x) = (1 - c_3 - c_4)x + c_3 x^2 + c_4 x^3.$$

In order to substitute into the functional, we compute the derivatives

$$\overline{y}' = 1 - c_3 - c_4 + 2c_3 x + 3c_4 x^2,$$

and

$$\overline{y}'' = 2c_3 + 6c_4 x.$$

For convenience of the substitution, we gather terms as

$$\overline{y}(x) = x + c_3(x^2 - x) + c_4(x^3 - x).$$

from which two basis functions to be used as weights are emerging as

$$w_1 = x^2 - x, w_2 = x^3 - x.$$

The Euler-Lagrange differential equation of the problem is

$$\overline{y}'' - x - 1 = 0,$$

hence Galerkin's integral equations become

$$\int_0^1 w_i(\overline{y}'' - x - 1)dx = 0; i = 1, 2.$$

Specifically for w_1

$$\int_0^1 (x^2 - x)\left((x + c_3(x^2 - x) + c_4(x^3 - x))'' - x - 1\right) dx = 0,$$

and for w_2

$$\int_0^1 (x^3 - x)\left((x + c_3(x^2 - x) + c_4(x^3 - x))'' - x - 1\right) dx = 0.$$

Rather involved gathering and integrating produce the system of equations

$$\frac{1}{3}c_3 + \frac{1}{2}c_4 - \frac{1}{4} = 0,$$

and

$$\frac{1}{2}c_3 + \frac{4}{5}c_4 - \frac{23}{60} = 0.$$

The solutions for the two unknown coefficients are

$$c_3 = \frac{1}{2},$$

and

$$c_4 = \frac{1}{6}.$$

With these, the Galerkin's approximate solution becomes

$$\overline{y}(x) = (1 - \frac{1}{2} - \frac{1}{6})x + \frac{1}{2}x^2 + \frac{1}{6}x^3,$$

which is finally

$$\overline{y}(x) = \frac{1}{3}x + \frac{1}{2}x^2 + \frac{1}{6}x^3.$$

This solution is identical to the Euler-Lagrange differential equation based analytic solution presented in Section 1.1, but this is a consequence of the simplicity of the problem. This is not generally true, but that does not diminish the usefulness of the method.

7.4   Approximate solutions of Poisson's equation

We will compare the prior two methods by solving Poisson's equation in its variational form by Ritz's method and in its differential equation form by Galerkin's.

The second order boundary value problem of Poisson's, introduced earlier, is presented in the variational form of

$$I(y) = \int\int_D \left(\left(\frac{\partial u}{\partial x}\right)^2 + \left(\frac{\partial u}{\partial y}\right)^2 + 2f(x,y)u(x,y) \right) dxdy$$

whose Euler-Lagrange equation leads to the form

$$\frac{\partial^2 u}{\partial x^2} + \frac{\partial^2 u}{\partial y^2} = f(x,y).$$

For the simplicity of the discussion and without loss of generality, we impose the boundary condition of

$$u = 0$$

on the perimeter of the domain D. Ritz's method indicates the use of the basis functions

$$b_i(x, y)$$

and demands that they also vanish on the boundary. The approximate solution in this two-dimensional case is

$$\overline{u}(x, y) = \sum_{i=1}^{n} \alpha_i b_i(x, y).$$

The partial derivatives are

$$\frac{\partial \overline{u}}{\partial x} = \sum_{i=1}^{n} \alpha_i \frac{\partial b_i(x, y)}{\partial x},$$

and

$$\frac{\partial \overline{u}}{\partial y} = \sum_{i=1}^{n} \alpha_i \frac{\partial b_i(x, y)}{\partial y}.$$

Substituting the approximate solution into the functional yields

$$I(\overline{u}) = \int \int_D \left(\left(\frac{\partial \overline{u}}{\partial x} \right)^2 + \left(\frac{\partial \overline{u}}{\partial y} \right)^2 + 2f(x, y)\overline{u}(x, y) \right) dx dy.$$

Evaluating the derivatives, this becomes

$$I(\overline{u}) = \int \int_D \left(\left(\sum_{i=1}^{n} \alpha_i \frac{\partial b_i}{\partial x} \right)^2 + \left(\sum_{i=1}^{n} \alpha_i \frac{\partial b_i}{\partial y} \right)^2 + 2f(x, y) \sum_{i=1}^{n} \alpha_i b_i \right) dx dy,$$

which may be reordered into the form

$$I(\overline{u}) = \sum_{i=1}^{n} \sum_{j=1}^{n} c_{ij} \alpha_i \alpha_j + 2 \sum_{i=1}^{n} d_i \alpha_i.$$

The coefficients are

$$c_{ij} = \int \int_D \left(\frac{\partial b_i}{\partial x} \frac{\partial b_j}{\partial x} + \frac{\partial b_i}{\partial y} \frac{\partial b_j}{\partial y} \right) dx dy$$

and

$$d_i = \int \int_D f(x, y) b_i dx dy.$$

As above, the unknown coefficients are solved from the conditions

$$\frac{\partial I(\overline{u})}{\partial \alpha_i} = 0, i = 1, 2, \ldots, n,$$

resulting in the linear system of equations

$$\sum_{j=1}^{n} c_{ij}\alpha_j + d_j = 0, i = 1, 2, \ldots, n.$$

It may be shown that the system is non-singular and always yields a non-trivial solution assuming that the basis functions form a linearly independent set. The computation of the terms of the equations, however, is rather tedious and resulted in the emergence of Galerkin's method.

Now we use Galerkin's method to solve Poisson's equation:

$$L(u) = \frac{\partial^2 u}{\partial x^2} + \frac{\partial^2 u}{\partial y^2} - f(x, y) = 0.$$

For this, Galerkin's method is presented as

$$\int\int_D \left(\frac{\partial^2 \overline{u}}{\partial x^2} + \frac{\partial^2 \overline{u}}{\partial y^2} - f(x, y) \right) b_j dx dy = 0, j = 1, \ldots, n.$$

Therefore

$$\int\int_D \left(\sum_{i=1}^{n} \alpha_i \frac{\partial^2 b_i}{\partial x^2} + \sum_{i=1}^{n} \alpha_i \frac{\partial^2 b_i}{\partial y^2} - f(x, y) \right) b_j dx dy = 0, j = 1, \ldots, n.$$

Reordering yields

$$\sum_{i=1}^{n} \alpha_i \int\int_D \left(\frac{\partial^2 b_i}{\partial x^2} + \frac{\partial^2 b_i}{\partial y^2} \right) b_j dx dy - \int\int_D f(x, y) b_j dx dy = 0, j = 1, \ldots, n.$$

The system of equations becomes

$$B\underline{a} = \underline{b}$$

with solution vector of

$$\underline{a} = \begin{bmatrix} \alpha_1 \\ \alpha_2 \\ \ldots \\ \alpha_n \end{bmatrix}.$$

The system matrix is of the form

$$B = \begin{bmatrix} B_{1,1} & B_{1,2} & \ldots & B_{1,n} \\ B_{2,1} & B_{2,2} & \ldots & B_{2,n} \\ \ldots & \ldots & \ldots & \ldots \\ B_{n,1} & B_{n,2} & \ldots & B_{n,n} \end{bmatrix}$$

whose terms are defined as

$$B_{j,i} = \int \int_D \left(\frac{\partial^2 b_i}{\partial x^2} + \frac{\partial^2 b_i}{\partial y^2} \right) b_j dx dy.$$

Finally, the right-hand side vector is

$$\underline{b} = \begin{bmatrix} \int \int_D f(x,y) b_1 dx dy \\ \int \int_D f(x,y) b_2 dx dy \\ \cdots \\ \int \int_D f(x,y) b_n dx dy \end{bmatrix}.$$

The comparison demonstrated that ultimately both methods result in solving linear systems and such systems are efficiently solved by readily available software tools in industrial applications.

While Galerkin's method also uses simple power function bases, it enables pre-computing some of the matrix components. This is an advantage over the Ritz method and provides a computational simplicity that is an important component in practice.

7.5   Kantorovich's method

Both the Ritz and Galerkin methods are restricted in their choices of basis functions, because their basis functions are required to satisfy the boundary conditions. The method of Kantorovich, described in [10], relaxes this restriction and enables the use of simpler basis functions.

Consider the variational problem of two variables

$$I(u) = \text{extremum}, (x,y) \in D,$$

with boundary conditions

$$u(x,y) = 0, (x,y) \in \partial D.$$

Here ∂D again denotes the boundary of the domain.

The method proposes the construction of a function ω, such that

$$\omega(x,y) \geq 0, (x,y) \in D,$$

and

$$\omega(x,y) = 0, (x,y) \in \partial D.$$

This function assumes the role of enforcing the boundary condition and the following set of simpler, power functions based, basis functions are adequate to present the solution:

$$b_1(x, y) = \omega(x, y),$$
$$b_2(x, y) = \omega(x, y)x,$$
$$b_3(x, y) = \omega(x, y)y,$$
$$b_4(x, y) = \omega(x, y)x^2,$$
$$b_5(x, y) = \omega(x, y)xy,$$
$$b_6(x, y) = \omega(x, y)y^2,$$

and so on, following the same pattern. It is clear that all these basis functions vanish on the boundary by the virtue of $\omega(x, y)$, even though the power function components do not.

The question is how to generate $\omega(x, y)$ for various shapes of domains. For a centrally located circle with radius r, the equation

$$x^2 + y^2 = r^2$$

implies very intuitively the form of

$$\omega(x, y) = r^2 - x^2 - y^2.$$

Obviously, the function is zero everywhere on the circle and non-zero in the interior of the domain. It is also non-zero on the outside of the domain, but that is irrelevant in connection with our problem.

One can also consider a domain consisting of overlapping circular regions, some of which represent voids in the domain. Figure 7.2 shows a domain of two circles with equations

$$x^2 + y^2 = r^2$$

and

$$(x - r/2)^2 + y^2 = (r/2)^2.$$

Reordering the latter yields

$$x^2 - xr + y^2 = 0,$$

and in turn results in

$$\omega(x, y) = (r^2 - x^2 - y^2)(x^2 - rx + y^2).$$

Clearly on the boundary of the larger circle, the left term is zero and on the boundary of the smaller circle, the right term is zero. Hence the product

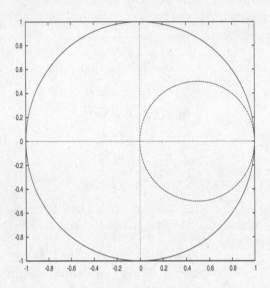

FIGURE 7.2 Domain with overlapping circular regions

function vanishes on the perimeter of both circles, which constitutes the now non-trivial boundary.

Let us now consider the boundary of a rectangle of width $2w$ and height $2h$, also centrally located around the origin. The equations of the sides

$$x = \pm w,$$

and

$$y = \pm h,$$

imply the very simple form of

$$\omega(x, y) = (w^2 - x^2)(h^2 - y^2).$$

The verification is very simple,

$$\omega(x, y) = 0; (x, y) = (\pm w, \pm h).$$

The construction technique clearly shows signs of difficulties to come with very generic, and especially three-dimensional domains. In fact such difficulties limited the practical usefulness of this otherwise innovative method until more recent work enabled the automatic creation of the ω functions for generic two- or three-dimensional domains with the help of spline functions, a topic which will be discussed in Chapter 9 at length.

We shall now demonstrate the correctness of such a solution. For this, we consider the solution of a specific Poisson's equation:

$$\frac{\partial^2 u}{\partial x^2} + \frac{\partial^2 u}{\partial y^2} = -2,$$

with

$$u(x, y) = 0, (x, y) \in \partial D,$$

where we designate the domain to be the rectangle whose ω function was specified above. We will search for Kantorovich's solution in the form of

$$u(x, y) = (w^2 - x^2)(h^2 - y^2)(\alpha_1 + \alpha_2 x + \alpha_3 y + \ldots).$$

Since the method is approximate, we may truncate the sequence of power function terms at a certain order. It is sufficient for the demonstration to use only the first term.

We will apply the method in connection with Galerkin's method of the last section. Therefore, the extremum is sought from

$$\int_{-w}^{+w} \int_{-h}^{+h} \left(\frac{\partial^2 u}{\partial x^2} + \frac{\partial^2 u}{\partial y^2} + 2 \right) \omega(x, y) dy dx = 0.$$

Executing the posted differentiations and substituting results in

$$\int_{-w}^{+w} \int_{-h}^{+h} -2\alpha_1 (w^2 - x^2)(h^2 - y^2)^2 - 2\alpha_1 (w^2 - x^2)^2 (h^2 - y^2) +$$

$$2(w^2 - x^2)(h^2 - y^2) dy dx = 0.$$

Since we only have a single coefficient, the system of equations developed earlier boils down to a single scalar equation of

$$b\alpha_1 = f,$$

with

$$b = \int_{-w}^{+w} \int_{-h}^{+h} \left((w^2 - x^2)(h^2 - y^2)^2 + (w^2 - x^2)^2 (h^2 - y^2) \right) dy dx,$$

and

$$f = \int_{-w}^{+w} \int_{-h}^{+h} (w^2 - x^2)(h^2 - y^2) dy dx.$$

After the (tedious) evaluation of the integrals, the value of

$$\alpha_1 = \frac{5}{4(w^2 + h^2)}$$

emerges. In turn, the approximate Kantorovich-Galerkin solution is

$$u(x,y) = \frac{5}{4} \frac{(w^2 - x^2)(h^2 - y^2)}{w^2 + h^2}.$$

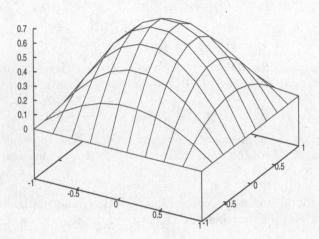

FIGURE 7.3 Solution of Poisson's equation

The solution is depicted graphically in Figure 7.3 using

$$w = h = 1.$$

The figure demonstrates that the solution function satisfies the zero boundary condition on the circumference of the square. To increase accuracy, additional terms of the power series may be used. The method also enables the exploitation of the symmetry of the domain. For example, if the above domain would exhibit the same height as width,

$$s = w = h,$$

the solution may be sought in the form of

$$u(x,y) = (s^2 - x^2)(s^2 - y^2)\left(\alpha_1 + \alpha_{23}(x+y)\right),$$

where α_{23} denotes the single constant accompanying both the second and third terms.

A generalization of this approach is necessary to eliminate the difficulties of producing an analytic ω function for practical domains with convoluted boundaries. The idea is to use an approximate solution to generate the ω function as well.

Let us consider the two-dimensional domain case and generate a surface approximation over the domain in the form of

$$\omega(x,y) = \sum_{i=0}^{n} \sum_{j=0}^{m} C_{i,j} B_i(x) B_j(y),$$

where the two sets of B basis functions are of common form, but dependent on either of the independent variables. The coefficients $C_{i,j}$ are either sampling points of the domain, or control points used to generate the surface. The latter case applies mainly to the interior points, and the earlier to the boundary.

This requires a simple Cartesian discretization of the domain along topological (possibly even equidistant) lines. The B-spline fitting technique introduced in Chapter 9 will provide the means for accomplishing this.

7.6   Boundary integral method

The boundary integral method is related to Kantorovich's method in the sense that both make use of the boundary-interior distinction of a variational problem. We will discuss this method in connection with a two-dimensional variational problem; however, the technique and conclusions apply to three dimensions as well. Let us consider the problem of

$$L(x,y)u(x,y) = f(x,y),$$

where $L(x,y)$ is a two-dimensional linear operator and the problem is defined on the domain $(x,y) \in \Omega$. The domain's boundary is Γ and the outward normal of the boundary, n, is defined.

The boundary integral method finds the solution in the form of

$$u(x,y) = \int_{\Omega} G(P,Q) f(x,y) d\Omega.$$

Here $G(P, Q)$ is Green's function corresponding to the particular linear operator. It is defined in terms of two points, $P = (x_p, y_p), Q = (x_q, y_q)$ as

$$L\left(G(P, Q)\right) = \delta(P - Q),$$

where δ is the Dirac function. Let us work with the two-dimensional Poisson's equation of the form

$$\Delta u(x, y) = f(x, y).$$

Here $L(x, y) = \Delta = \nabla^2$ and its Green's function is

$$G(P, Q) = \frac{1}{2\pi} \ln(r),$$

where

$$r = \sqrt{(x_p - x_q)^2 + (y_p - y_q)^2}.$$

The generic form of Green's theorem (a consequence of Gauss' divergence theorem) may be written as

$$\int_\Omega (u\nabla^2 v - v\nabla^2 u)d\Omega = \int_\Gamma (u\frac{\partial v}{\partial n} - v\frac{\partial u}{\partial n})d\Gamma.$$

Using Green's function in place of v we obtain

$$\int_\Omega (u\nabla^2 G - G\nabla^2 u)d\Omega = \int_\Gamma (u\frac{\partial G}{\partial n} - G\frac{\partial u}{\partial n})d\Gamma.$$

By definition

$$L\left(G(P, Q)\right) = \nabla^2 G(P, Q) = \delta(P - Q),$$

and due to the characteristics of the Dirac function, the first term on the left-hand side reduces to $u(x, y)$. Substituting the original equation into the second term, the resulting boundary integral solution becomes

$$u(x, y) = \int_\Omega Gf(x, y)d\Omega + \int_\Gamma u\frac{\partial G}{\partial n}d\Gamma - \int_\Gamma G\frac{\partial u}{\partial n}d\Gamma.$$

The first term on the right-hand side is the applied load in the domain and it is zero when the homogeneous Laplace problem is solved. The second term contains the Dirichlet boundary conditions via given boundary values of the function. The third term represents the Neumann boundary conditions by given derivatives with respect to the normal. It is possible that both types are given at the same time.

Assuming that the set of discretized points on the boundary are $q_j, j = 1, ..m$ and boundary conditions are given, the solution at any point in the interior may be computed as

$$u(x,y) = \int_\Omega Gf(x,y)d\Omega + \sum_{j=1}^m u(x_{q_j}, y_{q_j}) \int_{\Gamma_j} \frac{\partial G(p, q_j)}{\partial n} d\Gamma_j -$$

$$- \sum_{j=1}^m \frac{\partial u(x_{q_j}, y_{q_j})}{\partial n} \int_{\Gamma_j} G(p, q_j)d\Gamma_j.$$

Here the boundary segments are assigned to the given boundary points as

$$\Gamma = \sum_{j=1}^{m-1} \Gamma_j.$$

It is also possible to produce a discretized solution at a set of given interior points $p_i, i = 1, ...n$. In this case, a matrix formulation is possible (using the homogeneous case for simplicity of the presentation) as

$$u(x_{p_i}, y_{p_i}) = \sum_{j=1}^m A_{i,j} u(x_{q_j}, y_{q_j}) - \sum_{j=1}^m B_{i,j} \frac{\partial u(x_{q_j}, y_{q_j})}{\partial n},$$

where the matrix coefficients contain the pre-computed integrals

$$A_{i,j} = \int_{\Gamma_j} \frac{\partial G(p_i, q_j)}{\partial n} d\Gamma_j,$$

and

$$B_{i,j} = \int_{\Gamma_j} G(p_i, q_j)d\Gamma_j.$$

Let us now gather the solution points into the array

$$\underline{u} = \begin{bmatrix} u(x_{p_1}, y_{p_1}) \\ u(x_{p_2}, y_{p_2}) \\ ... \\ u(x_{p_n}, y_{p_n}) \end{bmatrix}.$$

Then the solution may be written as a simple matrix equation:

$$\underline{u} = A\underline{v} - B\underline{t},$$

where the vector containing the boundary condition displacement values is

$$\underline{v} = \begin{bmatrix} u(x_{q_1}, y_{q_1}) \\ u(x_{q_2}, y_{q_2}) \\ ... \\ u(x_{q_m}, y_{q_m}) \end{bmatrix}$$

and the vector holding the tangents is

$$\underline{t} = \begin{bmatrix} \frac{\partial u(x_{q_1},y_{q_1})}{\partial n} \\ \frac{\partial u(x_{q_2},y_{q_2})}{\partial n} \\ \cdots \\ \frac{\partial u(x_{q_m},y_{q_m})}{\partial n} \end{bmatrix}.$$

This is the approach of software tools using the boundary element method. The method is of engineering importance when the solution of a problem in the interior is largely homogeneous and the important solution variation is at or close to the boundary.

Let us now consider the case when only boundary tangents (Neumann boundary conditions) are given. Then the unknowns may be both in the interior and on the boundary as

$$u(x_{p_i}, y_{p_i}) = u(x_{q_j}, y_{q_j}),$$

when $i = j$. By the definition, the Green's function for the Laplace operator is singular when the solution point coincides with a boundary condition point and the solution integrals become improper. Hence the evaluation of the matrix coefficients must deal with that issue.

Nevertheless, the problem can be reformulated as

$$\sum_{j=1}^{m}(A_{i,j} + \frac{1}{2}\delta_{ij})u(x_{q_j},y_{q_j}) = \sum_{j=1}^{m} B_{i,j}\frac{\partial u(x_{q_j},y_{q_j})}{\partial n},$$

where δ_{ij} is the Kronecker delta. The problem is then of the form

$$\overline{A}\underline{u} = B\underline{t}.$$

Since the matrix on the left-hand side is now square, the system of equations may be formally solved as

$$\underline{u} = \overline{A}^{-1}B\underline{t}.$$

The singularity of the integrals carries into the system matrix by making it numerically ill-conditioned and requiring specialized solution techniques that avoid computing an explicit inverse.

The method is easily generalized to the three-dimensional Laplace operator whose Green's function is of the form

$$G(P,Q) = \frac{-1}{4\pi r},$$

where

$$r = \sqrt{x^2 + y^2 + z^2}.$$

Finding the Green's function for other operators is also possible. For example, the Green's function for the operator

$$L(x, y, t) = \partial_t^2 - \nabla^2$$

is defined also in terms of the Dirac function and the radius r as

$$G(P, Q) = \frac{\delta(t - r)}{4\pi r}.$$

This is the so-called d'Alembert operator of the wave equation that will be the subject of a mechanical problem (the elastic string) in Section 11.1.

7.7   Finite element method

The finite element method is an extension of Ritz's and Galerkin's approaches. Hence it is sometimes called the Ritz-Galerkin finite element method.

The method of finite elements extends the basis function approximation concept with a subdivision or discretization of the interval of interest. This extension enables the use of lower order basis functions than those of Ritz or Galerkin, in many cases linear interpolation. On the other hand, the result will be at the discrete locations of the interval and not an approximate continuous function as in Ritz's or Galerkin's methods.

The finite element method solves the variational problem

$$I = \int_{x_1}^{x_n} f(x, y, y') dx = \text{extremum}$$

by using approximate solutions in a collection of segments of the interval

$$I = \sum_{i=1}^{n-1} \int_{x_i}^{x_{i+1}} f\left(x, \overline{y}_i(x), \overline{y}_i'(x)\right) dx = \text{extremum}.$$

The linear interpolation basis functions for the segments are of the form

$$\overline{y}_i(x) = y_i \frac{x_{i+1} - x}{x_{i+1} - x_i} + y_{i+1} \frac{x - x_i}{x_{i+1} - x_i}.$$

The derivatives become

$$\overline{y}_i'(x) = y_i \frac{-1}{x_{i+1} - x_i} + y_{i+1} \frac{1}{x_{i+1} - x_i} = \frac{y_{i+1} - y_i}{x_{i+1} - x_i}.$$

The y_i discrete scalar values at the end points of the segments are the unknown values of the solution function at the corresponding locations. They are the solutions of the system of equations

$$\frac{\partial I}{\partial y_i} = 0; i = 2, ..., n - 1.$$

Note that the interval end points are not included as they are the boundary conditions satisfied by the interpolations in the first and the last segments. The intermediate values are not exactly in agreement with the analytic solution, hence the method is approximate. To obtain solution values inside the segments, the linear interpolation function that is valid for that segment is used. As such, it is again an approximation of the analytic solution.

To illustrate this conceptually different method, let us view the example that we also solved via Galerkin's method:

$$I = \int_0^1 \left(\frac{1}{2} y'^2 + (x + 1)y \right) dx = \text{extremum}$$

with boundary conditions

$$y(0) = 0, y(1) = 1.$$

We will use three interior segments to discretize the interval as

$$x_1 = 0, x_2 = \frac{1}{3}, x_3 = \frac{2}{3}, x_4 = 1.$$

Hence there will be three solution function sections:

$$\overline{y}_1(x) = y_1 \frac{x_2 - x}{x_2 - x_1} + y_2 \frac{x - x_1}{x_2 - x_1}; \ 0 \le x \le \frac{1}{3},$$

$$\overline{y}_2(x) = y_2 \frac{x_3 - x}{x_3 - x_2} + y_3 \frac{x - x_2}{x_3 - x_2}; \ \frac{1}{3} \le x \le \frac{2}{3},$$

and

$$\overline{y}_3(x) = y_3 \frac{x_4 - x}{x_4 - x_3} + y_4 \frac{x - x_3}{x_4 - x_3}; \ \frac{2}{3} \le x \le 1.$$

Similarly the derivatives are computed in three sections:

$$\overline{y}_1' = \frac{y_2 - y_1}{x_2 - x_1}; \ 0 \le x \le \frac{1}{3},$$

$$\overline{y}_2' = \frac{y_3 - y_2}{x_3 - x_2}; \ \frac{1}{3} \le x \le \frac{2}{3},$$

and

$$\overline{y}_3' = \frac{y_4 - y_3}{x_4 - x_3}; \ \frac{2}{3} \le x \le 1.$$

Substitution into the functional, integrating and summing we obtain

$$I = 3y_2^2 + 3y_3^2 - 3y_2 y_3 + \frac{4}{9}y_2 - \frac{22}{9}y_3 + \frac{49}{7}.$$

The derivative based system of equations becomes

$$\frac{\partial I}{\partial y_2} = 6y_2 - 3y_3 + \frac{4}{9} = 0,$$

and

$$\frac{\partial I}{\partial y_3} = -3y_2 + 6y_3 - \frac{22}{9} = 0.$$

The solution is

$$y_2 = \frac{14}{81}, y_3 = \frac{40}{81},$$

along with the dictated boundary conditions

$$y_1 = 0, y_4 = 1.$$

Evaluating the solution at the mid-point of the interval we obtain

$$\overline{y}\left(\frac{1}{2}\right) = \overline{y}_2\left(\frac{1}{2}\right) = \frac{14}{81}\frac{2/3 - 1/2}{2/3 - 1/3} + \frac{40}{81}\frac{1/2 - 1/3}{2/3 - 1/3} = \frac{34}{81}.$$

To put the accuracy into perspective, we compare it to the analytic solution computed in Section 1.1 as

$$y = \frac{1}{6}x^3 + \frac{1}{2}x^2 + \frac{1}{3}x.$$

The exact solution value at the mid-point is

$$y\left(\frac{1}{2}\right) = \frac{15}{48},$$

and the error of the approximate solution is

$$\frac{139}{1296}$$

or approximately 0.1. The discussion here was intended to illustrate the concept of the discretized and interpolated solution approach of finite elements. The practical importance of the method is in continuum mechanical problems of two and three dimensions, the detailed subject of Chapter 12.

7.8   Exercises

Compare your results with the analytic solution given.

1.
Use Euler's method to find an approximate solution. Use 5 intermediate segments.
$I = \int_0^1 (2y + y'^2)dx = $ extremum.
Boundary conditions $y(0) = 0, y(1) = 1$.

2.
Use Ritz's method to find an approximate solution. Use power function bases.
$I = \int_0^1 (xy' + y'^2)dx = $ extremum.
Boundary conditions $y(0) = 0, y(1) = 1$.

3.
Use Galerkin's method to find an approximate solution. Use power function bases.
$I = \int_0^1 (4xy' + y'^2)dx = $ extremum.
Boundary conditions $y(0) = 0, y(1) = 1$.

4. Use the finite element method to find an approximate solution. Use three segments in the interval.
$\int_0^1 (y'^2 + 12xy)dx = $ extremum.
Boundary conditions $y(0) = 0, y(1) = 1$.

5.
Use any numerical method to find an approximate solution.
$\int_0^1 (y'')^2 dx = $ extremum.
Boundary conditions $y(0) = 0, y(1) = 1, y'(0) = 0, y'(1) = 3$.

Part II

Modeling applications

8

Differential geometry

Differential geometry is a classical mathematical area that has become very important for engineering applications in the recent decades. This importance is based on the rise of computer-aided visualization and geometry generation technologies.

The chapter will address the fundamental problem of differential geometry, the finding of geodesic curves, that has practical implications in manufacturing. Development of non-mathematical surfaces used in ships and airplanes has serious financial impact in reducing material waste and improving the quality of the surfaces.

While the discussion in this chapter will focus on analytically solvable problems, the methods and concepts we introduce will provide a foundation applicable in various engineering areas.

8.1 The geodesic problem

Finding a geodesic curve on a surface is a classical problem of differential geometry. Variational calculus seems uniquely applicable to this problem. Let us consider a parametrically given surface

$$\underline{r} = x(u,v)\underline{i} + y(u,v)\underline{j} + z(u,v)\underline{k}.$$

Let two points on the surface be

$$\underline{r}_0 = x(u_0,v_0)\underline{i} + y(u_0,v_0)\underline{j} + z(u_0,v_0)\underline{k},$$

and

$$\underline{r}_1 = x(u_1,v_1)\underline{i} + y(u_1,v_1)\underline{j} + z(u_1,v_1)\underline{k}.$$

The shortest path on the surface between these two points is the **geodesic curve**. Consider the square of the arc length

$$ds^2 = (dx)^2 + (dy)^2 + (dz)^2,$$

and compute the differentials related to the parameters,

$$ds^2 = E(u,v)(du)^2 + 2F(u,v)dudv + G(u,v)(dv)^2.$$

Here the so-called first fundamental quantities are defined as

$$E(u,v) = \left(\frac{\partial x}{\partial u}\right)^2 + \left(\frac{\partial y}{\partial u}\right)^2 + \left(\frac{\partial z}{\partial u}\right)^2 = |\underline{r}'_u|^2,$$

$$F(u,v) = \frac{\partial x}{\partial u}\frac{\partial x}{\partial v} + \frac{\partial y}{\partial u}\frac{\partial y}{\partial v} + \frac{\partial z}{\partial u}\frac{\partial z}{\partial v} = \underline{r}'_u \cdot \underline{r}'_v,$$

and

$$G(u,v) = \left(\frac{\partial x}{\partial v}\right)^2 + \left(\frac{\partial y}{\partial v}\right)^2 + \left(\frac{\partial z}{\partial v}\right)^2 = |\underline{r}'_v|^2.$$

Assume that the equation of the geodesic curve in the parametric space is described by

$$v = v(u).$$

Then the geodesic curve is the solution of the variational problem

$$I(v) = \int_{u_0}^{u_1} \sqrt{E(u,v) + 2F(u,v)\frac{dv}{du} + G(u,v)(\frac{dv}{du})^2}\, du = \text{extremum}$$

with boundary conditions

$$v(u_0) = v_0,$$

and

$$v(u_1) = v_1.$$

The corresponding Euler-Lagrange differential equation is

$$\frac{E_v + 2v'F_v + v'^2 G_v}{2\sqrt{E(u,v) + 2F(u,v)v' + G(u,v)v'^2}} - \frac{d}{du}\frac{F + Gv'}{\sqrt{E(u,v) + 2F(u,v)v' + G(u,v)v'^2}} = 0,$$

with the notation of

$$E_v = \frac{\partial E}{\partial v}, F_v = \frac{\partial F}{\partial v}, G_v = \frac{\partial G}{\partial v},$$

and

$$v' = \frac{dv}{du}.$$

The equation is rather difficult in general, and exploitation of special cases arising from the particular surface definitions is advised.

When the first fundamental quantities are only functions of the u parameter, the equation simplifies to

$$\frac{F + Gv'}{\sqrt{E(u,v) + 2F(u,v)v' + G(u,v)v'^2}} = c_1.$$

A further simplification is based on the practical case when the u and v parametric lines defining the surface are orthogonal. In this case

$$F = 0,$$

and the equation may easily be integrated as

$$v = c_1 \int \frac{\sqrt{E}}{\sqrt{G^2 - c_1^2 G}} du + c_2.$$

The integration constants may be resolved from the boundary conditions.

When the first fundamental quantities are only functions of the v parameter, and the $F = 0$ assumption still holds, the equation becomes

$$\frac{Gv'^2}{\sqrt{E + Gv'^2}} - \sqrt{E + Gv'^2} = c_1.$$

Reordering and another integration bring

$$v = c_1 \int \frac{\sqrt{E^2 - c_1^2 E}}{\sqrt{G}} dv + c_2.$$

8.1.1 Geodesics of a sphere

For an enlightening example, we consider a sphere, given by

$$x(u,v) = R\sin(v)\cos(u),$$

$$y(u,v) = R\sin(v)\sin(u),$$

and

$$z(u,v) = R\cos(v).$$

The first fundamental quantities encapsulating the surface information are

$$E = R^2 \sin^2(v),$$

$$F = 0,$$

and

$$G = R^2.$$

Since this is the special case consisting of only v, the equation of the geodesic curve becomes

$$u = c_1 \int \frac{R}{\sqrt{R^4 \sin^4(v) - c_1^2 R^2 \sin^2(v)}} dv + c_2.$$

After the integration by substitution and some algebraic manipulations, we get

$$u = -a \sin \frac{\cot(v)}{\sqrt{\left(\frac{R}{c_1}\right)^2 - 1}} + c_2.$$

It follows that

$$\sin(c_2)\,(R\sin(v)\cos(u)) - \cos(c_2)\,(R\sin(v)\sin(u)) - \frac{R\cos(v)}{\sqrt{\left(\frac{R}{c_1}\right)^2 - 1}} = 0.$$

Substituting the surface definition of the sphere yields

$$x\,\sin(c_2) - y\,\cos(c_2) - \frac{z}{\sqrt{\left(\frac{R}{c_1}\right)^2 - 1}} = 0$$

and that represents an intersection of the sphere with a plane. By substituting boundary conditions, it would be easy to see that the actual plane contains the origin and defines the great circle going through the two given points. This fact is manifested in everyday practice by the transoceanic airplane routes' well-known northern swing in the Northern Hemisphere.

8.1.2   Geodesic polyhedra

Let us consider a case of three intersecting geodesics of a sphere and the resulting spherical triangle that is called an Euler triangle. Any such spherical triangle has angles α, β and γ at the three corners. If the radius of the sphere is r, then the area of the spherical triangle is

$$A = r^2[(\alpha + \beta + \gamma) - \pi].$$

The term in the bracket is called the spherical excess, essentially describing the difference between the area of the planar triangle spanning the three inter-section points and that of the spherical triangle. This excess could be rather significant.

Let us consider the unit sphere and the main circles of the $x - z$, $x - y$ and $y - z$ planes' intersections with the sphere. Focusing only on the first octant of the sphere, the angles at the intersection of each pair of these circles is 90 degrees, resulting in an excess of also 90 degrees, or $\pi/2$. The area of that spherical triangle according to the above formula is $1^2 \cdot \pi/2$, which is one-eighth of the surface of the unit sphere.

The three intersection points, the corners of the spherical triangle are $(1,0,0), (0,1,0)$, and $(0,0,1)$. The distances between any pair of these points is $\sqrt{2}$, hence the area of the planar triangle is $\frac{\sqrt{3}}{2}$, which is significantly less than that of the spherical triangle.

The triangularization of a sphere is a seamless covering the whole surface of a sphere with a set of connecting, but not overlapping spherical triangles. If the spherical triangles on the sphere are replaced by the corresponding planar triangles, the result is the tessellated sphere, which is a geodesic polyhedron, shown in Figure 8.1.

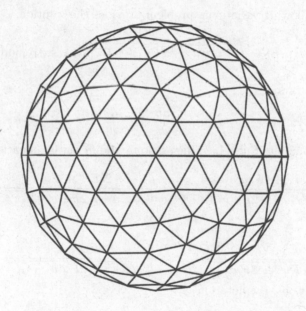

FIGURE 8.1 Tessellated sphere

Tessellation, or generating geodesic polyhedra, is an important component of the surface meshing technology of the finite element technique, described in more detail in Section 12.1.1.

8.2   A system of differential equations for geodesic curves

Let us now seek the geodesic curve in the parametric form of

$$u = u(t),$$

and

$$v = v(t).$$

The curve goes through two points

$$P_0 = (u(t_0), v(t_0)),$$

and

$$P_1 = (u(t_1), v(t_1)).$$

Then the following variational problem provides the solution:

$$I(u, v) = \int_{t_0}^{t_1} \sqrt{Eu'^2 + 2Fu'v' + Gv'^2}\, dt = \text{extremum}.$$

Here

$$u' = \frac{du}{dt}, v' = \frac{dv}{dt}.$$

The corresponding Euler-Lagrange system of differential equations is

$$\frac{E_u u'^2 + 2F_u u'v' + G_u v'^2}{2\sqrt{Eu'^2 + 2Fu'v' + Gv'^2}} - \frac{d}{dt}\frac{2(Eu' + Fv')}{\sqrt{Eu'^2 + 2Fu'v' + Gv'^2}} = 0,$$

and

$$\frac{E_v u'^2 + 2F_v u'v' + G_v v'^2}{2\sqrt{Eu'^2 + 2Fu'v' + Gv'^2}} - \frac{d}{dt}\frac{2(Fu' + Gv')}{\sqrt{Eu'^2 + 2Fu'v' + Gv'^2}} = 0.$$

In the equations, the notation

$$E_u = \frac{\partial E}{\partial u}, \ F_u = \frac{\partial F}{\partial u}, \ G_u = \frac{\partial G}{\partial u}$$

was used.

A more convenient and practically useful formulation, without the explicit derivatives, based on [5] is

$$u'' + \Gamma_{11}^1 u'^2 + 2\Gamma_{12}^1 u'v' + \Gamma_{22}^1 v'^2 = 0,$$

and

$$v'' + \Gamma_{11}^2 u'^2 + 2\Gamma_{12}^2 u'v' + \Gamma_{22}^2 v'^2 = 0.$$

Here

$$u'' = \frac{d^2 u}{dt^2}, v'' = \frac{d^2 v}{dt^2}.$$

The Γ are the Christoffel symbols that are defined as

$$\Gamma_{11}^1 = \frac{GE_u - 2FF_u + FE_v}{2(EG - F^2)},$$

$$\Gamma_{12}^1 = \frac{GE_v - FG_u}{2(EG - F^2)},$$

$$\Gamma_{22}^1 = \frac{2GF_v - GG_u - FG_v}{2(EG - F^2)},$$

$$\Gamma_{11}^2 = \frac{2EF_u - EE_v - FE_u}{2(EG - F^2)},$$

$$\Gamma_{12}^2 = \frac{EG_u - FE_v}{2(EG - F^2)},$$

and

$$\Gamma_{22}^2 = \frac{EG_v - 2FF_v + FG_u}{2(EG - F^2)}.$$

These formulae all require that

$$EG - F^2 \neq 0$$

which is true when a parameterization is regular.

8.2.1   Geodesics of surfaces of revolution

Another practically important special case is represented by surfaces of revolution. Their generic description may be of the form

$$x = u \cos(v),$$

$$y = u \sin(v),$$

and

$$z = f(u).$$

Here the last equation describes the meridian curve generating the surface. The first order fundamental terms are

$$E = 1 + f'^2(u),$$

$$F = 0,$$

and

$$G = u^2.$$

The solution following the discussion in Section 8.1 becomes

$$v = c_1 \int \frac{\sqrt{1 + f'^2(u)}}{u\sqrt{u^2 - c_1^2}} du + c_2.$$

A simple sub-case of this class is a unit cylinder, described as

$$x = \cos(v),$$
$$y = \sin(v),$$

and

$$z = u.$$

The geometric meaning of the v parameter is the rotation angle generating the cylinder's circumference and the u parameter is the axial direction of the surface. The fundamental terms are

$$E = 1,$$
$$F = 0,$$

and

$$G = 1.$$

The equation of the geodesic curve on the cylinder following above is

$$v = c_1 \int \frac{1}{1\sqrt{1 - c_1^2}} du + c_2,$$

or

$$v = c_1 \frac{1}{\sqrt{1 - c_1^2}} \int du + c_2.$$

With

$$c_3 = c_1 \frac{1}{\sqrt{1 - c_1^2}}$$

and integration we obtain

$$v = c_3 u + c_2.$$

In the general case, this is a helix on the surface of the cylinder going through the two points. This is also a line in the parametric space. This fact is geometrically easy to explain because the cylinder is a developable surface. Such a surface may be rectified onto a plane. In such a case, the geodesic curve is a straight line on the rectifying plane. The only curvature of the helix will be that of the cylinder.

The constants of integration may be determined from the boundary conditions. For example, assume the case shown in Figure 8.2, where the starting

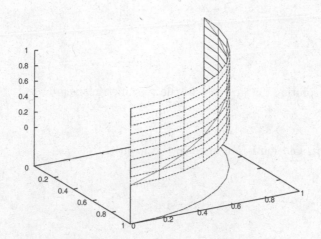

FIGURE 8.2 Geodesic curve of a cylinder

point is at

$$P_0 = (x_0, y_0, z_0) = (1, 0, 0)$$

corresponding to parametric coordinates

$$u(t_0) = 0, v(t_0) = 0.$$

The endpoint is located at

$$P_1 = (x_1, y_1, z_1) = (0, 1, 1)$$

corresponding to parametric coordinates

$$u(t_1) = 1, v(t_1) = \pi/2.$$

Substituting the starting point yields

$$0 = c_3 \cdot 0 + c_2,$$

which results in

$$c_2 = 0.$$

The endpoint substitution produces

$$\pi/2 = c_3 \cdot 1 + c_2,$$

and in turn

$$c_3 = \frac{\pi}{2}.$$

The specific solution for this case in the parametric space is

$$v = \frac{\pi}{2}u.$$

The Cartesian solution is obtained in the form of

$$x = \cos(v),$$
$$y = \sin(v),$$

and

$$z = \frac{v}{\pi/2}.$$

It is easy to see that the last equation makes the elevation change from zero to 1, in accordance with the turning of the helix.

Since the parametric space of the cylinder is simply the rectangle of the developed surface, it is easy to see some special sub-cases. If the two points are located at the same rotational position (v=constant), but at different heights, the geodesic curve is a straight line. If the two points are on the same height (u=constant), but at different rotational angles, the geodesic curve is a circular arc.

The last two sections demonstrated the difficulties of finding the geodesic curves even on regular surfaces like the sphere or the cylinder. On a general three-dimensional surface, these difficulties increase significantly and may render using the differential equation of the geodesic curve unfeasible.

8.3   Geodesic curvature

Let us consider the parametric curve

$$\underline{r}(t) = x(t)\underline{i} + y(t)\underline{j} + z(t)\underline{k}$$

on the surface

$$S(u,v) = x(u,v)\underline{i} + y(u,v)\underline{j} + z(u,v)\underline{k}.$$

Let $\underline{n}$ denote the unit normal of the surface. The curvature vector of a three-dimensional curve is defined as

$$\underline{k} = \frac{d\underline{t}}{dt} = \underline{t}',$$

where $\underline{t}$ is the tangent vector computed as

$$\underline{t} = \frac{d\underline{r}}{dt},$$

and also assumed to be a unit vector (a unit speed curve) for the simplicity of the derivation. Then the unit bi-normal vector is

$$\underline{b} = \underline{n} \times \underline{t}.$$

We represent the curvature vector with components along the bi-normal vector and the normal vector $\underline{n}$ at any point as

$$\underline{k} = \kappa_n \underline{n} + \kappa_g \underline{b}.$$

The coefficients are the normal curvature and the geodesic curvature, respectively. Taking the inner product of the last equation with the $\underline{b}$ vector and exploiting the perpendicularity conditions present, we obtain

$$\underline{b} \cdot \underline{k} = \kappa_g.$$

Substituting the definition of the bi-normal and the curvature vector results in

$$\kappa_g = (\underline{n} \times \underline{t}) \cdot \underline{t}'.$$

For the more generic case when the tangent and normal vectors are not of unit length, the geodesic curvature of a curve is defined as

$$\kappa_g = \frac{\underline{r}''(t) \cdot (\underline{n} \times \underline{r}'(t))}{\left\|\underline{r}'(t)\right\|^3}.$$

A curve on a surface is called geodesic if at each point of the curve its principal normal and the surface normal are collinear. Therefore:

A curve $\underline{r}(t)$ on the surface $S(u,v)$ is geodesic if the geodesic curvature of the curve is zero.

In order to prove that, the terms are computed from the surface information, such as

$$\underline{r}' = \underline{t} = \frac{\partial f}{\partial u} u' + \frac{\partial f}{\partial v} v' = f_u u' + f_v v'.$$

The application of the chain rule results in

$$\underline{r}'' = \underline{t}' = f_{uu}(u')^2 + 2f_{uv}u'v' + f_{vv}(v')^2 + f_u u'' + f_v v''.$$

After substitution into the equation of the geodesic curvature and some algebraic work, while employing again the Christoffel symbols, [5] produces the form

$$\kappa_g = \sqrt{EG - F^2}\Big(\Gamma_{11}^2(u')^3 + (2\Gamma_{12}^1 - \Gamma_{11}^1)(u')^2 v' +$$

$$(\Gamma_{22}^2 - 2\Gamma_{12}^1)u'(v')^2 - \Gamma_{22}^1(v')^3 + u'v'' - u''v'\Big).$$

Since

$$EG - F^2 \neq 0,$$

the term in the brackets must be zero for zero geodesic curvature. That happens when the following terms vanish

$$u'' + \Gamma_{11}^1 u'^2 + 2\Gamma_{12}^1 u'v' + \Gamma_{22}^1 v'^2 = 0,$$

and

$$v'' + \Gamma_{11}^2 u'^2 + 2\Gamma_{12}^2 u'v' + \Gamma_{22}^2 v'^2 = 0.$$

This result is the decoupled system of equations of the geodesic, introduced in Section 8.1, hence the vanishing of the geodesic curvature is indeed a characteristic of a geodesic curve.

Finally, since the recent discussions were mainly on parametric forms, the equation of the geodesic for an explicitly given surface

$$z = z(x, y(x))$$

is quoted from [5] for completeness' sake without derivation:

$$\left(1 + \left(\frac{\partial z}{\partial x}\right)^2 + \left(\frac{\partial z}{\partial y}\right)^2\right)\frac{d^2 y}{dx^2} = \frac{\partial z}{\partial x}\frac{\partial^2 z}{\partial y^2}\left(\frac{dy}{dx}\right)^3 +$$

$$\left(2\frac{\partial z}{\partial x}\frac{\partial^2 z}{\partial x \partial y} - \frac{\partial z}{\partial y}\frac{\partial^2 z}{\partial y^2}\right)\left(\frac{dy}{dx}\right)^2 +$$

$$\left(\frac{\partial z}{\partial x}\frac{\partial^2 z}{\partial x^2} - 2\frac{\partial z}{\partial y}\frac{\partial^2 z}{\partial x \partial y}\right)\frac{dy}{dx} - \frac{\partial z}{\partial y}\frac{\partial^2 z}{\partial x^2}.$$

The formula is rather overwhelming and useful only in connection with the simplest surfaces.

8.3.1 Geodesic curvature of helix

Let us enlighten this further by reconsidering the case of the geodesic curve of the cylinder discussed in the last section. The geodesic curve we obtained was the helix:

$$r(t) = \cos(t)\underline{i} + \sin(t)\underline{j} + \frac{t}{\pi/2}\underline{k}.$$

The appropriate derivatives are

$$r'(t) = -\sin(t)\underline{i} + \cos(t)\underline{j} + \frac{1}{\pi/2}\underline{k}$$

and

$$r''(t) = -\cos(t)\underline{i} - \sin(t)\underline{j} + 0\underline{k}.$$

The surface normal is computed as

$$\underline{n} = \frac{\partial S}{\partial u} \times \frac{\partial S}{\partial v}.$$

In the specific case of the cylinder

$$S(u, v) = \cos(v)\underline{i} + \sin(v)\underline{j} + u\underline{k},$$

it becomes

$$\underline{n} = \cos(t)\underline{i} + \sin(t)\underline{j} + 0\underline{k}.$$

The cross-product and substitution of $v = t$ yields

$$\underline{n}(t) \times r'(t) = \frac{2}{\pi}\sin(t)\underline{i} - \frac{2}{\pi}\cos(t)\underline{j} + \left(\sin^2(t) + \cos^2(t)\right)\underline{k}.$$

The numerator of the curvature becomes zero, as

$$(-\cos(t)\underline{i} - \sin(t)\underline{j} + 0\underline{k}) \cdot \left(\frac{2}{\pi}\sin(t)\underline{i} - \frac{2}{\pi}\cos(t)\underline{j} + \left(\sin^2(t) + \cos^2(t)\right)\underline{k}\right) = 0.$$

Since the denominator

$$\left\| r'(t) \right\|^3 = \left(\sqrt{1 + \left(\frac{1}{\pi/2}\right)^2} \right)^3$$

is not zero, the geodesic curvature becomes zero. Hence, the helix is truly the geodesic curve of the cylinder.

The concept of geodesic curves may be generalized to spaces of higher dimensions. The geodesic curve notation, however, while justified on a three-dimensional surface, needs to be generalized as well. In such cases, one talks about finding a geodesic object, or just a geodesic, as opposed to a curve on a surface.

8.4   Generalization of the geodesic concept

Insofar, we confined the geodesic problem to finding a curve on a three-dimensional surface, but the concept may be generalized to higher dimensions. Physicists use the space-time continuum as a four-dimensional (Minkowski) space and find geodesic paths in that space. The arc length definition of this space is

$$ds^2 = dx^2 + dy^2 + dz^2 - cdt^2,$$

where t is the time dimension and c is the speed of light. The variational problem of minimal arc length may be posed similarly as in Section 8.1 and may be solved with similar techniques. Einstein used this generalization to explain planetary motion as a geodesic phenomenon in the four-dimensional space.

9

Computational geometry

The geodesic concept, introduced in the last chapter purely on variational principles, has interesting engineering aspects. On the other hand, the analytic solution of a geodesic curve by finding the extremum of a variational problem may not be easy in practical cases.

The subject of this chapter is to provide a computational representation of various geometry objects also from a variational foundation. The technology of splines, natural and constrained, will be explored and extended to surfaces and even volumes. Their practical use will also be described.

9.1 Natural splines

It is reasonable to assume that the quality of a curve in a geodesic sense is related to its curvature. This observation proposes a strategy for creating good quality (albeit not necessarily geodesic) curves by minimizing the curvature.

Since the curvature is difficult to compute, one can use the second derivative of the curve in lieu of the curvature. This results in the following variational problem statement for a smooth curve: Find the curve $s(t)$ that results in

$$I(s) = \int_{t_0}^{t_n} k(s'')^2 dt = \text{extremum}.$$

The constant k represents the fact that this is an approximation of the curvature, but will be left out from our work below. This variational problem leads to the well-known spline functions.

Let us consider the following variational problem. Find the curve between two points P_0, P_3 such that

$$I(s) = \int_{t_0}^{t_3} \left(\frac{d^2 s}{dt^2}\right)^2 dt = \text{extremum},$$

with boundary conditions of

$$P_0 = s(t_0), P_3 = s(t_3),$$

and additional discrete internal constraints of

$$P_1 = s(t_1), P_2 = s(t_2).$$

In essence, we are constraining two interior points of the curve, along with the fixed beginning and endpoints. We will, for the sake of simplicity, assume unit equidistant parameter values as

$$t_i = i, i = 0. \ldots, 3.$$

While the functional does not contain the independent variable t and the dependent variable $s(t)$, it is of higher order. Hence, the Euler-Poisson equation of second order applies:

$$\frac{\partial f}{\partial y} - \frac{d}{dx}\frac{\partial f}{\partial y'} + \frac{d^2}{dx^2}\frac{\partial f}{\partial y''} = 0,$$

and in this case it simplifies to

$$\frac{d^4}{dt^4} s(t) = 0.$$

Straightforward integration yields the solution of the form

$$s(t) = c_0 + c_1 t + c_2 t^2 + c_3 t^3,$$

where c_i are constants of integration to be satisfied by the boundary conditions. Imposing the boundary conditions and constraints yields the system of equations

$$\begin{bmatrix} 1 & 0 & 0 & 0 \\ 1 & 1 & 1 & 1 \\ 1 & 2 & 4 & 8 \\ 1 & 3 & 9 & 27 \end{bmatrix} \begin{bmatrix} c_0 \\ c_1 \\ c_2 \\ c_3 \end{bmatrix} = \begin{bmatrix} P_0 \\ P_1 \\ P_2 \\ P_3 \end{bmatrix}.$$

The inversion of the system matrix results in the generating matrix

$$M = \begin{bmatrix} 1 & 0 & 0 & 0 \\ -11/6 & 3 & -3/2 & 1/3 \\ 1 & -5/2 & 2 & -1/2 \\ -1/6 & 1/2 & -1/2 & 1/6 \end{bmatrix}$$

for the natural spline. For any given set of four points

$$P = \begin{bmatrix} x_0 & y_0 & z_0 \\ x_1 & y_1 & z_1 \\ x_2 & y_2 & z_2 \\ x_3 & y_3 & z_3 \end{bmatrix}$$

the coefficients of the solution may be obtained by

$$C = MP,$$

with distinct coefficients for all coordinate directions as

$$C = \begin{bmatrix} c_0^x & c_0^y & c_0^z \\ c_1^x & c_1^y & c_1^z \\ c_2^x & c_2^y & c_2^z \\ c_3^x & c_3^y & c_3^z \end{bmatrix}.$$

For example, the points

$$P = \begin{bmatrix} 1 & 1 \\ 2 & 2 \\ 3 & 2 \\ 4 & 3 \end{bmatrix}$$

result in coefficients

$$C = \begin{bmatrix} 1 & 1 \\ 1 & 13/6 \\ 0 & -3/2 \\ 0 & 1/3 \end{bmatrix}.$$

The parametric solution curve is of the form

$$x(t) = 1 + t,$$

$$y(t) = 1 + 13/6t - 3/2t^2 + 1/3t^3.$$

The example solution curve is shown in Figure 9.1, where the input points are connected by the straight lines that represent the chords of the spline. The spline demonstrates a good smoothness while satisfying the constraints.

Several extensions of this problem are noteworthy. It is possible to pose the variational problem in two-parameter form as

$$I(s) = \int \int_D \left(\left(\frac{\partial}{\partial u} s(u, v) \right)^2 + \left(\frac{\partial}{\partial v} s(u, v) \right)^2 \right) du dv = \text{extremum}.$$

The Euler-Lagrange equation corresponding to this problem arrives at Laplace's equation:

$$\frac{\partial^2}{\partial u^2} s(u, v) + \frac{\partial^2}{\partial v^2} s(u, v) = 0.$$

This is sometimes called the harmonic equation, hence the splines so obtained are also called harmonic splines.

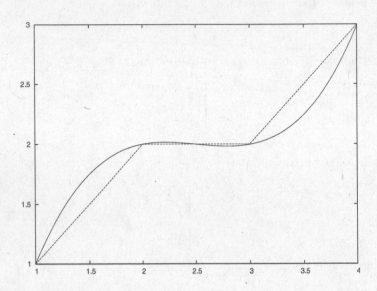

FIGURE 9.1 Natural spline approximation

It is also often the case in practice that many more than four points are given:

$$P_i = (x_i, y_i, z_i), i = 0, \ldots, n.$$

This enables the generation of a multitude of natural spline segments, and a curvature continuity condition between the segments may also be enforced. Finally, the direct (for example Ritz) solution of the above variational problem leads to the widely used B-splines, a topic of the next chapter.

9.2   B-spline approximation

As shown in Chapter 7, when using numerical methods, an approximate solution is sought in terms of suitable basis functions:

$$\overline{s}(t) = \sum_{i=0}^{n} B_{i,k}(t) Q_i,$$

where Q_i are the yet unknown control points (i=0,...,n) and $B_{i,k}$ are the basis functions of degree k in the parameter t. For industrial computational work, the class of basis functions resulting in the so-called B-splines are defined in [1] as

$$B_{i,0}(t) = \begin{cases} 1, t_i \le t < t_{i+1} \\ 0, t < t_i, t \ge t_{i+1} \end{cases}$$

where the higher order terms are recursively computed:

$$B_{i,k}(t) = \frac{t - t_i}{t_{i+k} - t_i} B_{i,k-1}(t) + \frac{t_{i+k+1} - t}{t_{i+k+1} - t_{i+1}} B_{i+1,k-1}(t).$$

The basis functions are computed from specific parameter values, called knot values. If their distribution is not equidistant, the splines are called non-uniform B-splines. If they are uniformly distributed, they are generating uniform B-splines.

The knot values are a subset of the parameter space and their selection enables a unique control of the behavior of the spline. For example, the use of duplicate knot values inside the parameters span of the spline results in a local change. The use of multiple knot values at the boundaries enforces various end conditions, such as the frequently used clamped end condition. This control mechanism contributes to the popularity of the method in computer-aided design (CAD), but will not be further explored here.

Smoothing a B-spline is defined by the variational problem

$$I_s(Q) = \int_{t_0}^{t_n} \left(\sum_{i=0}^{n} B''_{i,k}(t) Q_i \right)^2 dt = \text{extremum}.$$

The derivative of the basis functions may be recursively computed. For $k = 1$, since $B_{i,0}$ are constant

$$\frac{d}{dt} B_{i,1}(t) = B'_{i,1}(t) = \frac{1}{t_{i+1} - t_i} B_{i,0}(t) - \frac{1}{t_{i+2} - t_{i+1}} B_{i+1,0}(t).$$

For $k = 2$

$$\frac{d}{dt} B_{i,2}(t) = B'_{i,2}(t) = \frac{1}{t_{i+2} - t_i} B_{i,1}(t) + \frac{t - t_i}{t_{i+2} - t_i} B'_{i,1}(t) -$$

$$\frac{1}{t_{i+3} - t_i} B_{i+1,1}(t) + \frac{t_{i+3} - t}{t_{i+3} - t_{i+1}} B'_{i+1,1}(t).$$

Substituting the $k = 1$ derivative into the second term results in

$$\frac{t - t_i}{t_{i+2} - t_i} B'_{i,1}(t) = \frac{t - t_i}{t_{i+2} - t_i} \left(\frac{1}{t_{i+1} - t_i} B_{i,0}(t) - \frac{1}{t_{i+2} - t_{i+1}} B_{i+1,0}(t) \right) =$$

$$\frac{1}{t_{i+2} - t_i} \left(\frac{t - t_i}{t_{i+1} - t_i} B_{i,0}(t) + \frac{t_i - t}{t_{i+2} - t_{i+1}} B_{i+1,0}(t) \right).$$

The content of the parenthesis is easily recognizable as $B_{i,1}(t)$, hence this term is identical to the first. Similar arithmetic on the second two terms results in the derivative for $k = 2$ as

$$\frac{d}{dt} B_{i,2}(t) = B'_{i,2}(t) = \frac{2}{t_{i+2} - t_i} B_{i,1} - \frac{2}{t_{i+3} - t_{i+1}} B_{i+1,1}.$$

By induction, for any k value, the first derivative is as follows:

$$\frac{d}{dt} B_{i,k}(t) = B'_{i,k}(t) = \frac{k}{t_{i+k} - t_i} B_{i,k-1}(t) - \frac{k}{t_{i+k+1} - t_{i+1}} B_{i+1,k-1}(t).$$

A repeated application of the same step will produce the needed second derivative B'' as

$$\frac{d}{dt} B'_{i,k}(t) = B''_{i,k}(t) = \frac{k}{t_{i+k} - t_i} B'_{i,k-1}(t) - \frac{k}{t_{i+k+1} - t_{i+1}} B'_{i+1,k-1}(t).$$

The spline, besides being smooth (minimal in curvature), is expected to approximate a given set of points $P_j; j = 0, \ldots, m$, with associated prescribed parameter values (not necessarily identical to the knot values) of $t_j; j = 0, \ldots, m$. If such parameter values are not given, the parameterization may be via the simple method of uniform spacing defined as $t_j = j; 0 \leq j \leq m$. Assuming that the point set defined is geometrically semi-equidistant, this is proven in industry to be a good method for the problem at hand. If that is not the case, a parameterization based on the chord length may also be used.

Approximating the given points with the spline is another variational problem that requires finding a minimum of the squares of the distances between the spline and the points.

$$I_a(\overline{s}) = \sum_{j=0}^{m} \left(\overline{s}(t_j) - P_j \right)^2 = \text{extremum}.$$

Substituting the B-spline formulation and the basis functions results in

$$I_a(Q) = \sum_{j=0}^{m} \left(\sum_{i=0}^{n} B_{i,k}(t_j) Q_i - P_j \right)^2.$$

Similarly, in the smoothing variational problem, we also replace the integral with a sum over the given points in the parameter span, resulting in

$$I_s(Q) = \sum_{j=0}^{m} \left(\sum_{i=0}^{n} B''_{i,k}(t_j) Q_i \right)^2.$$

Finally, the functional to produce a smooth spline approximation is the sum of the two functionals

$$I(Q) = I_a(Q) + I_s(Q).$$

The notation is to demonstrate the dependence upon the yet unknown control points of the spline.

According to the numerical method developed in Chapter 7, the problem has an extremum when

$$\frac{\partial I}{\partial Q_i} = 0,$$

for each $i = 0, \ldots, n$. The control points will be, of course, described by Cartesian coordinates; hence, each of the above equations represents three scalar equations.

The derivative of the approximating component with respect to an unknown control point Q_p yields

$$\frac{\partial I_a}{\partial Q_p} - 2 \sum_{j=0}^{m} B_{p,k}(t_j) \left(\sum_{i=0}^{n} B_{i,k}(t_j) Q_i - P_j \right) = 0,$$

where $p = 0, 1, \ldots, n$. This is expressed in matrix form as

$$B^T B Q - B^T P$$

with the matrices

$$B = \begin{bmatrix} B_{0,k}(t_0) & B_{1,k}(t_0) & B_{2,k}(t_0) & \ldots & B_{n,k}(t_0) \\ B_{0,k}(t_1) & B_{1,k}(t_1) & B_{2,k}(t_1) & \ldots & B_{n,k}(t_1) \\ \ldots & \ldots & \ldots & \ldots & \ldots \\ B_{0,k}(t_m) & B_{1,k}(t_m) & B_{2,k}(t_m) & \ldots & B_{n,k}(t_m) \end{bmatrix},$$

$$P = \begin{bmatrix} P_0 \\ P_1 \\ \ldots \\ P_m \end{bmatrix},$$

and

$$Q = \begin{bmatrix} Q_0 \\ Q_1 \\ \ldots \\ Q_n \end{bmatrix}.$$

For degree $k = 3$, the basis functions may be analytically computed as:

$$B_{0,3} = \frac{1}{6}(1 - t)^3,$$

$$B_{1,3} = \frac{1}{6}(3t^3 - 6t^2 + 4),$$

$$B_{2,3} = \frac{1}{6}(-3t^3 + 3t^2 + 3t + 1),$$

and

$$B_{3,3} = \frac{1}{6}t^3.$$

Furthermore, for uniform parameterization, the B matrix is easily computed by hand as

$$B = \frac{1}{6}\begin{bmatrix} 1 & 4 & 1 & 0 \\ 0 & 1 & 4 & 1 \\ -1 & 4 & -5 & 8 \\ -8 & 31 & -44 & 27 \\ -27 & 100 & -131 & 64 \end{bmatrix}.$$

The derivative of the smoothing component of the functional, with respect to an unknown control point Q_p yields

$$\frac{\partial I_s}{\partial Q_p} = 2\sum_{j=1}^{m} B''_{p,k}(t_j)\sum_{i=0}^{n} B''_{i,k}(t_j)Q_i = 0,$$

where $p \in [0,\ldots,n]$. This results in a smoothing matrix

$$D = \begin{bmatrix} B''_{0,k}(t_0) & B''_{1,k}(t_0) & \ldots & B''_{n,k}(t_0) \\ B''_{0,k}(t_1) & B''_{1,k}(t_1) & \ldots & B''_{n,k}(t_1) \\ \ldots & \ldots & \ldots & \ldots \\ B''_{0,k}(t_m) & B''_{1,k}(t_m) & \ldots & B''_{n,k}(t_m) \end{bmatrix}.$$

These second derivatives for the cubic case are

$$B''_{0,3} = 1 - t,$$

$$B''_{1,3} = 3t - 2,$$

$$B''_{2,3} = -3t + 1,$$

and

$$B''_{3,3} = t.$$

For uniform parameterization, the smoothing matrix is computed as

$$D = \frac{1}{6}\begin{bmatrix} 1 & -2 & 1 & 0 \\ 0 & 1 & -2 & 1 \\ -1 & 4 & -5 & 2 \\ -2 & 7 & -8 & 3 \\ -3 & 10 & -11 & 4 \end{bmatrix}.$$

The simultaneous solution of both the smoothing and approximating problem is now represented by the matrix equation

$$AQ = B^T P$$

where the system matrix is

$$A = B^T B + D^T D.$$

The solution of this system produces the control points for a smooth approximation.

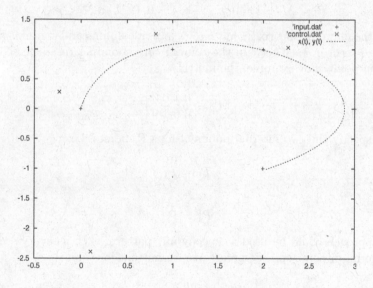

FIGURE 9.2 Smooth B-spline approximation

Figure 9.2 shows the smooth spline approximation for a set of given points. The input points as well as the computed control points are also shown. Note that, as opposed to the natural spline, the curve does not go through the points exactly, but it is very smooth.

9.3   B-splines with point constraints

It is possible to require some of the points to be exactly satisfied. For the generic case of multiple enforced points, a constrained variational problem is formed.

$$I(Q_i) = \text{extremum},$$

subject to

$$\bar{s}(t_l) = R_l; l = 0, 1, \ldots, o.$$

Here the enforced points R_l represent a subset of the given points (P_j), while the remainder are to be approximated. The subset is specified as

$$R_l = M P_j; l = 0, 1, \ldots, o; j = 0, 1, \ldots, m; o < m,$$

where the mapping matrix M has o rows and m columns and contains a single non-zero term in each row, in the column corresponding to a selected interpolation point. For example, the matrix

$$M = \begin{bmatrix} 0\ 1\ 0\ 0 \\ 0\ 0\ 1\ 0 \end{bmatrix}$$

would specify the two internal points of four P_j points, i.e.,

$$R_0 = P_1$$

and

$$R_1 = P_2.$$

This approach could be used to specify any pattern, such as every second or third term, or some specific points at intermittent locations.

Introducing the specifics of the splines and Lagrange multipliers, the constrained variational problem is presented as

$$I(Q_i, \lambda_l) = I(Q_i) + \sum_{l=0}^{o} \lambda_l \left(\sum_{i=0}^{n} (B_{i,k}(t_l)Q_i) - R_l \right).$$

The derivatives of $I(Q_i)$ with respect to the Q_p control point were computed earlier, but need to be extended with the derivative of the term containing the Lagrange multiplier:

$$\sum_{l=0}^{o} B_{p,k}(t_l)\lambda_l \sum_{i=0}^{n} B_{i,k}(t_l).$$

Utilizing the earlier introduced matrices, this term is

$$B^T M^T \Lambda,$$

where Λ is a column vector of $o + 1$ Lagrange multipliers,

$$\begin{bmatrix} \lambda_0 \\ \lambda_1 \\ \cdots \\ \lambda_o \end{bmatrix}.$$

The derivatives with respect to the Lagrange multipliers λ_l produce

$$\frac{\partial I(Q_i, \lambda_l)}{\partial \lambda_l} = \sum_{i=0}^{n} (B_{i,k}(t_l)Q_i) - R_l = 0; l = 0, 1, \ldots, o.$$

This results in $o + 1$ new equations of the form

$$\sum_{i=0}^{n} B_{i,k}(t_l)Q_i = R_l,$$

or in matrix form, using the earlier matrices:

$$MBQ = R,$$

where R is a vector of the interpolated points. The two sets of equations may be assembled into a single matrix equation with $n+1+o+1$ rows and columns of the form

$$\begin{bmatrix} A & B^T M^T \\ MB & 0 \end{bmatrix} \begin{bmatrix} Q \\ \Lambda \end{bmatrix} = \begin{bmatrix} B^T P \\ MP \end{bmatrix}.$$

The first matrix row represents the constrained functional and the second row represents the constraints. The simultaneous solution of this (symmetric, indefinite, but still linear) system produces the optimized (approximated and smoothed) and selectively interpolated solution.

The solution of this problem is accomplished in the following steps. First, the unknown control points are expressed from the first row of

$$AQ + B^T M^T \Lambda = B^T P$$

as functions of the unknown Lagrange multipliers. Substituting into the second equation is the way to compute the multipliers:

$$\Lambda = (MBA^{-1}B^T M^T)^{-1}(MBA^{-1}B^T P - MP).$$

Finally, the set of control points, which are solutions of the constrained variational problem, are obtained explicitly from the first equation as

$$Q = A^{-1}(B^T P - B^T M^T \Lambda).$$

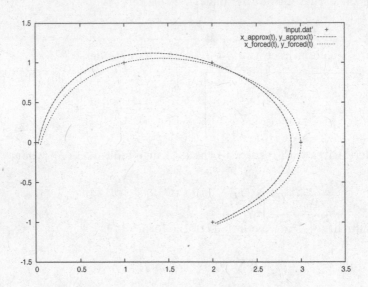

FIGURE 9.3 Point constrained B-spline

The effect of point constraints is shown in Figure 9.3 in connection with
the earlier example, constraining the spline to go through the second and the
fourth points. The dashed curve is the original curve, while the dotted curve
is the new curve and it demonstrates the adherence to the constraint, at the
same time maintaining the quality of the approximation and the smoothness.

9.4   B-splines with tangent constraints

It may be desirable for the engineer to be able to enforce constraints posed
by specifying tangents at certain points. These are of the form

$$\overline{s}'(t_l) = T_l; l = 0, 1, \ldots, o,$$

assuming for now that the tangents are given at the same points where interpolation constraints were also given. The constrained problem shown in the
prior section will be extended with the additional constraints and Lagrange
multipliers:

$$I(Q_i, \lambda_l) = I(Q_i) + \sum_{l=0}^{o} \lambda_l \left(\sum_{i=0}^{n} (B'_{i,k}(t_l)Q_i) - T_l \right).$$

The derivatives with respect to the new Lagrange multipliers are

$$\frac{\partial I(Q_i, \lambda_l)}{\partial \lambda_l} = \sum_{i=0}^{n} (B'_{i,k}(t_l)Q_i) - T_l = 0; l = 0, 1, \ldots, o.$$

This results in $o + 1$ new equations of the form

$$\sum_{i=0}^{n} B'_{i,k}(t_l)Q_i = T_l,$$

or in matrix form, using some of the earlier matrices:

$$MCQ = T,$$

where T is a vector of the given tangents and the matrix of first derivatives is

$$C = \begin{bmatrix} B'_{0,k}(t_0) & B'_{1,k}(t_0) & \ldots & B'_{n,k}(t_0) \\ B'_{0,k}(t_1) & B'_{1,k}(t_1) & \ldots & B'_{n,k}(t_1) \\ \ldots & \ldots & \ldots & \ldots \\ B'_{0,k}(t_m) & B'_{1,k}(t_m) & \ldots & B'_{n,k}(t_m) \end{bmatrix}.$$

The first derivatives of the basis functions for the cubic case with uniform parametrization are

$$B'_{0,3} = -\frac{1}{2}(1 - t)^2,$$

$$B'_{1,3} = \frac{3}{2}t^2 - 2t,$$

$$B'_{2,3} = -\frac{3}{2}t^2 + t + \frac{1}{2},$$

and

$$B'_{3,3} = \frac{1}{2}t^2.$$

For the uniform case, the C matrix containing the first derivatives is

$$C = \frac{1}{2} \begin{bmatrix} -1 & 0 & 1 & 0 \\ 0 & -1 & 0 & 1 \\ -1 & 4 & -7 & 4 \\ -4 & 15 & -20 & 9 \\ -9 & 32 & -39 & 16 \end{bmatrix}.$$

The three sets of equations may be assembled into one matrix equation with $n + 1 + 2(o + 1)$ rows and columns of the form

$$\begin{bmatrix} A & B^T M^T & C^T M^T \\ MB & 0 & 0 \\ MC & 0 & 0 \end{bmatrix} \begin{bmatrix} Q \\ \Lambda_p \\ \Lambda_t \end{bmatrix} = \begin{bmatrix} B^T P \\ MP \\ MT \end{bmatrix}.$$

The index of the Lagrange multipliers refers to points (p) or tangents (t).

The restriction of giving tangents at all the same points where interpolation constraints are also given may be relaxed and the points with tangential constraints may only be a subset of the points where interpolation constraints are placed. In this case, the final equation is

$$\begin{bmatrix} A & B^T M^T & C^T N^T \\ MB & 0 & 0 \\ NC & 0 & 0 \end{bmatrix} \begin{bmatrix} Q \\ \Lambda_p \\ \Lambda_t \end{bmatrix} = \begin{bmatrix} B^T P \\ MP \\ NT \end{bmatrix}.$$

Here the N mapping matrix is a subset of the M mapping matrix. The solution of this problem is similar to the solution of the simply constrained case, albeit a bit more tedious, due to the fact that the constraints are now of two different kinds. First, the solution in terms of the multipliers is expressed

$$Q = A^{-1} B^T P - A^{-1} \begin{bmatrix} B^T M^T & C^T N^T \end{bmatrix} \begin{bmatrix} \Lambda_p \\ \Lambda_t \end{bmatrix}.$$

Then there is a matrix equation to compute the multipliers from

$$\begin{bmatrix} MB \\ NC \end{bmatrix} Q = \begin{bmatrix} MP \\ NT \end{bmatrix}.$$

The sets of multipliers are obtained from

$$\begin{bmatrix} \Lambda_p \\ \Lambda_t \end{bmatrix} = \left(\begin{bmatrix} MB \\ NC \end{bmatrix} A^{-1} \begin{bmatrix} B^T M^T & C^T N^T \end{bmatrix} \right)^{-1} \left(\begin{bmatrix} MB \\ NC \end{bmatrix} A^{-1} (B^T P) - \begin{bmatrix} MP \\ NT \end{bmatrix} \right)$$

by executing the posted matrix operations. Finally, the desired set of control points satisfying the constrained problem are

$$Q = A^{-1} (B^T P - B^T M^T \Lambda_p - C^T N^T \Lambda_t).$$

Let us use again the same set of points, but enforce the curve going through the second point with a tangent of 45 degrees.

The dotted curve in Figure 9.4 demonstrates the satisfaction of both constraints, going through the second point and having a 45-degree tangent. It is also very clear that such a strong constraint imposed upon one point has a significant effect on the shape of the curve, but the smoothness of the curve is still excellent.

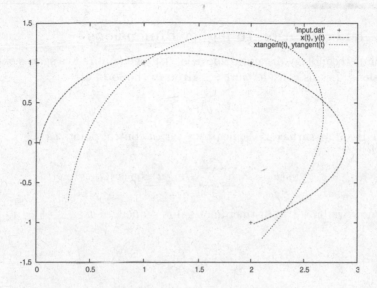

FIGURE 9.4 Tangent constrained B-spline

In practical applications, some heuristics, like setting the tangent at a certain point parallel to the chord between the two neighboring points, can be used successfully. Then

$$T_i = \frac{\overline{P_{i+1} - P_i}}{\left\| \overline{P_{i+1} - P_i} \right\|}.$$

This would result in different control points and a much smaller deformation of the overall curve may be obtained. Systematic and possibly interactive application of this concept should result in good shape preservation and general smoothness.

9.5   Generalization to higher dimensions

The spline technology discussed above is easily generalized to spaces of higher dimensions. Let us consider surfaces given in the form of

$$z(x,y) = f(x,y)$$

first. A B-spline surface is defined by a set of control points as

$$\overline{s}(u,v) = \sum_{i=0}^{n} \sum_{j=0}^{m} B_{i,k}(u) B_{j,k}(v) Q_{ij},$$

where now we have two distinct knot value sequences of

$$u_i; i = 0, 1, \ldots, n,$$

and

$$v_j; j = 0, 1, \ldots, m.$$

Here the k is again the order of the approximation that is usually 3. The rectangular arrangement of the control points is not necessary and may be overcome by coalescing certain points or adjustments of the knot points. The control points to provide a smooth approximation of the given geometric surface are selected from the variational problem of

$$I(\overline{s}) = \int \int \left(f(x,y) - \overline{s}(x,y) \right)^2 dx dy = \text{extremum}.$$

Substituting the surface spline definition and sampling of the given surface results in another, albeit more difficult, system of linear equations from which the control point locations may be resolved in a similar fashion as in the case of spline curves before.

Finally, it is also possible to describe some geometrical volumes with the B-spline technology. Consider the form

$$\overline{s}(u,v,t) = \sum_{i=0}^{n} \sum_{j=0}^{m} \sum_{l=0}^{p} B_{i,k}(u) B_{j,k}(v) B_{l,k}(t) Q_{ijl},$$

where now the third set of knot values

$$t_l; l = 0, 1, \ldots, p$$

defines the direction through the volume starting from a surface. The rectangular arrangement is applied again, but may be overcome with some inconvenience.

9.6 Weighting and non-uniform parametrization

It is also possible to generalize the B-spline technology by the use weights, resulting in rational B-splines. When a non-uniform parameterization is also used, the splines become Non-Uniform, Rational B-splines, known as NURBS [13].

Introducing weights associated with each control point results in the rational B-spline curve of form

$$S(t) = \frac{\sum_{i=0}^{n} w_i B_{i,k}(t) Q_i}{\sum_{i=0}^{n} w_i B_{i,k}(t)}.$$

The geometric meaning of the weights is simply to pull the curve closer to certain input points. It is, however, important to point out that changing one single weight value will result only in a local shape change in the segment related to the point. This local control is a spectacular advantage in modeling geometric objects.

The other, not always recognized advantage of the weighted technology is the ability to exactly fit conic sections, such as circles or parabolas. That is not possible without the weights. One can ask why do this, since conic sections have simple closed form descriptions. The advantage lies in the seamless modeling process of a geometry dominated by non-mathematical curves and surfaces, but occasionally interspersed with conic sections. The whole model will be comprised of splines, some integer, some rational, but with simply enforced continuity between the segments.

The weighted, or rational, formulation extends quite easily to surfaces:

$$S(u,v) = \frac{\sum_{i=0}^{n} \sum_{j=0}^{m} w_{i,j} B_{i,k}(u) B_{j,l}(v) Q_{i,j}}{\sum_{i=0}^{n} \sum_{j=0}^{m} w_{i,j} B_{i,k}(u) B_{j,l}(v)}.$$

Note that the degree of the v directional parametric curve may be different than that of the u curve, denoted by l. Similarly, the parameterization in both directions may be different. This gives tremendous flexibility to the method.

Geometric modeling operations are enabled by these objects. Consider generating a swept surface by moving a curve $C(u)$ along a trajectory $T(v)$. This is conceptually similar to generating a cylinder by defining a circle and the axis perpendicular to the plane of the circle. In general, the surface generated by this process may be described as

$$S(u,v) = C(u) + T(v).$$

Assume that the curves are rational B-splines of the same order k,

$$C(u) = \frac{\sum_{i=0}^{n} w_i^C B_{i,k}(u) Q_i^C}{\sum_{i=0}^{n} w_i^C B_{i,k}(u)}$$

and

$$T(v) = \frac{\sum_{j=0}^{m} w_j^T B_{j,k}(v) Q_j^T}{\sum_{j=0}^{m} w_j^T B_{j,k}(v)}.$$

Then the swept rational B-spline surface is of form

$$S(u,v) = \frac{\sum_{i=0}^{n} \sum_{j=0}^{m} w_{i,j} B_{i,k}(u) B_{j,k}(v) Q_{i,j}}{\sum_{i=0}^{n} \sum_{j=0}^{m} w_{i,j} B_{i,k}(u) B_{j,k}(v)},$$

where

$$Q_{i,j} = Q_i^C + Q_j^T$$

and

$$w_{i,j} = w_i^C w_j^T.$$

The superscripts C and T stand for the curve and trajectory, respectively. Similar considerations may be used to generate rational B-spline surfaces of revolution around a given axis.

Finally, rational B-splines also generalize to three dimensions for modeling volumes:

$$S(u,v,t) = \frac{\sum_{i=0}^{n} \sum_{j=0}^{m} \sum_{p=0}^{q} w_{i,j,p} B_{i,k}(u) B_{j,k}(v) B_{p,k}(t) Q_{i,j,p}}{\sum_{i=0}^{n} \sum_{j=0}^{m} \sum_{p=0}^{q} w_{i,j,p} B_{i,k}(u) B_{j,k}(v) B_{p,k}(t)}.$$

The form is written with the assumption of the curve degree being the same (k) in all three parametric directions, albeit that is not necessary.

It is important to point out that the surface representations via B-splines may also produce non-rectangular surface patches. Such, for example triangular, patches are very important in the finite element discretization step to be discussed in Chapter 12. They may easily be produced from the above formulations by collapsing a pair of points into one.

9.7   Industrial applications

Both of the generalizations in the last two sections are important in computer-aided design (CAD) tools in the industry. They represent an efficient way to describe general (non-mathematical) surfaces and volumes.

Another, very important higher than three-dimensional extension occurs in computer-aided manufacturing (CAM). The calculation of smooth tool-paths of five-axis machines includes the three Cartesian coordinates and two additional quantities related to the tool position. This is important in machining of surface parts comprised of valleys and walls.

The positioning of the cutting tool is customarily described by two angles. The tool's "leaning" in the normal plane is one which may be construed as a rotation with respect to the bi-normal of the path curve. The tool's "swaying" from the normal plane, which constitutes a rotation around the path tangent as an axis, may be another angle.

Abrupt changes in the tool axes are detrimental to the machined surface quality as well as to the operational efficiency. Hence, it is a natural desire to smooth these quantities as well. The variational formulation for the geometric smoothing of the spline, shown above, accommodates any number of such additional considerations.

A recent application of the techniques developed in this chapter is in the area of 3D printing. A body printed is described by a B-spline volume described earlier. Any printing technology proceeds along a certain axis specifically related to the shape of the body. The contour curves of the cross-sections of the body perpendicular to the axis of printing are 2D B-spline curves. The printing area in each layer is bounded by these curves and controls the printing process.

10

Variational equations of motion

We encountered variational forms of equations of motion in prior chapters, for example, when solving the brachistochrone problem in Section 1.4.2. Several dynamic equations of motion will be derived from variational principles in this chapter using techniques developed by Legendre, Hamilton and Lagrange. Specifically, mechanical systems, electric circuits and orbital motion will be investigated.

10.1 Legendre's dual transformation

This transformation invented by Legendre is of fundamental importance. Let us consider the function of n variables

$$f = f(u_1, u_2, ..., u_n).$$

Legendre proposed to introduce a new set of variables by the transformation of

$$v_i = \frac{\partial f}{\partial u_i}; i = 1, 2, ..., n.$$

The Hessian matrix of this transformation is

$$H(f) = \begin{bmatrix} \frac{\partial^2 f}{\partial u_1^2} & \frac{\partial^2 f}{\partial u_1 \partial u_2} & \cdots & \frac{\partial^2 f}{\partial u_1 \partial u_n} \\ \frac{\partial^2 f}{\partial u_2 \partial u_1} & \frac{\partial^2 f}{\partial u_2^2} & \cdots & \frac{\partial^2 f}{\partial u_2 \partial u_n} \\ \cdots & \cdots & \cdots & \cdots \\ \frac{\partial^2 f}{\partial u_n \partial u_1} & \frac{\partial^2 f}{\partial u_n \partial u_2} & \cdots & \frac{\partial^2 f}{\partial u_n^2} \end{bmatrix}.$$

If the determinant of this matrix, sometimes called the Hessian, is not zero, then the variables of the new set are independent. This means that they could also be obtained as functions of the original variables.

We define a new function in terms of the new variables

$$g = g(v_1, v_2, ..., v_n).$$

The two functions are related as

$$g = \sum_{i=1}^{n} u_i v_i - f.$$

The notable consequence is the spectacular duality between the two sets. The original variables can be now expressed as

$$u_i = \frac{\partial g}{\partial v_i}; i = 1, 2, ..., n.$$

and the original function regained as

$$f = \sum_{i=1}^{n} u_i v_i - g.$$

Legendre's transformation is completely symmetric.

Let us now look at a function of two sets of variables:

$$f = f(u_1, u_2, ..., u_n, w_1, w_2, ..., w_n).$$

If the variables of the second set are independent of the first, they are considered to be parameters and the transformation will retain them as such:

$$g = g(v_1, v_2, ..., v_n, w_1, w_2, ..., w_n).$$

The relationship between the two functions regarding the parameters is

$$\frac{\partial f}{\partial w_i} = -\frac{\partial g}{\partial w_i}, i = 1, 2, ..., n.$$

This transformation will be instrumental when applied to the functions introduced in the next sections.

10.2 Hamilton's principle

Hamilton's principle is a generalization of Euler's principle of minimum action, introduced earlier in Section 1.4.4 in connection with the problem of a particle moving under the influence of a gravity field. Hamilton's principle, however, is much more general and it is applicable to complex mechanical systems. For conservative (energy preserving) systems, Hamilton's principle states that the motion between two points is on the path of least action defined by the variational problem of

$$\int_{t_1}^{t_2} L\,dt = \text{extremum},$$

with the Lagrangian function

$$L = E_k - E_p,$$

where E_k and E_p are the kinetic and potential energy, respectively. Hence the principle may also be stated as

$$\int_{t_1}^{t_2} (E_k - E_p)\,dt = \text{extremum},$$

where the extremum is not always zero. The advantageous feature of Hamilton's principle is that it is stated in terms of energies, which are independent of the selection of coordinate systems. Hamilton's principle is of fundamental importance because many of the general physical laws may be derived from it as we will see in the next sections.

10.3   Hamilton's canonical equations

Let us view the Lagrangian as a function of n generalized displacements and velocities, and time as

$$L = L(q_i, \dot{q}_i, t),$$

for $i = 1, 2, \ldots n$.

Hamilton's canonical equations are the result of the application of Legendre's transformation to the Lagrangian function. Specifically Hamilton transformed the velocity components as

$$p_i = \frac{\partial L}{\partial \dot{q}_i}.$$

Differentiating and applying the Euler-Lagrange differential equation, we obtain

$$\dot{p}_i = \frac{d}{dt}p_i = \frac{d}{dt}\frac{\partial L}{\partial \dot{q}_i} = \frac{\partial L}{\partial q_i}.$$

On the other hand, the total differential of the Lagrangian is

$$dL = \frac{\partial L}{\partial t}dt + \sum_{i=1}^{n}\left(\frac{\partial L}{\partial q_i}dq_i + \frac{\partial L}{\partial \dot{q}_i}d\dot{q}_i\right).$$

Substituting results in

$$dL = \frac{\partial L}{\partial t}dt + \sum_{i=1}^{n}(\dot{p}_i dq_i + p_i d\dot{q}_i).$$

Reordering yields

$$d(\sum_{i=1}^{n} p_i\dot{q}_i - L) = -\frac{\partial L}{\partial t}dt - \sum_{i=1}^{n}(\dot{p}_i dq_i - \dot{q}_i dp_i),$$

where we exploited the total differential

$$d(p_i\dot{q}_i) = dp_i\dot{q}_i + p_i d\dot{q}_i.$$

Introducing the function

$$H = \sum_{i=1}^{n}(p_i\dot{q}_i - L),$$

called the Hamiltonian that is now only a function of the new and old generalized displacement variables and time:

$$H = H(p_i, q_i, t).$$

Its total differential is

$$dH = \frac{\partial H}{\partial t}dt + \sum_{i=1}^{n}\left(\frac{\partial H}{\partial p_i}dp_i + \frac{\partial H}{\partial q_i}dq_i\right).$$

Matching terms between the dH and dL differentials produces the relationship

$$\frac{\partial H}{\partial t} = -\frac{\partial L}{\partial t}.$$

Hamilton's canonical equations are then

$$\dot{q}_i = \frac{\partial H}{\partial p_i},$$

and

$$\dot{p}_i = -\frac{\partial H}{\partial q_i},$$

for $i = 1, 2, \ldots n$. The p_i, q_i are called canonical variables. There are twice as many first order equations as number of components, but being first order, this system is easy to solve by matrix methods.

Legendre's duality is clearly present. The time variable is the parameter unchanged by the Legendre transformation and it satisfies the same equation derived in Section 10.1 as

$$\frac{\partial H}{\partial t} = -\frac{\partial L}{\partial t}.$$

The dual relationship of the Lagrangian and Hamiltonian functionals is also clearly satisfied

$$H = \sum_{i=1}^{n} p_i \dot{q}_i - L,$$

and

$$L = \sum_{i=1}^{n} p_i \dot{q}_i - H.$$

The double dimensional space of the canonical variables is called the phase-space, sometimes also called the $q - p$ space. When the time variable t is added, the space is called the state-space, an instrumental platform in structural mechanics.

10.3.1 Conservation of energy

The relationship between the two functionals is not always easy to establish. Let us consider conservative systems in which the potential energy is only a function of the displacement generalized variables as

$$E_p = E_p(q),$$

while the kinetic energy is a quadratic function of the derivative generalized variables (or generalized velocities):

$$E_k = E_k(\dot{q}_i^2).$$

Hence

$$2E_k = \sum_{i=1}^{n} \frac{\partial E_k}{\partial \dot{q}_i} \dot{q}_i.$$

Substituting the canonical variables from the last section, we obtain

$$2E_k = \sum_{i=1}^{n} \frac{\partial L}{\partial \dot{q}_i} \dot{q}_i = \sum_{i=1}^{n} p_i \dot{q}_i.$$

Therefore, the Hamiltonian becomes

$$H = \sum_{i=1}^{n} p_i \dot{q}_i - L = 2E_k - (E_k - E_p) = E_k + E_p.$$

This relationship states that the Hamiltonian is the sum of kinetic and potential energy. Let us now further investigate the Hamiltonian. Since it is of the form

$$H = H(q_1, q_2, ..., q_n, p_1, p_2, ..p_n),$$

its derivative with respect to time is

$$\frac{dH}{dt} = \sum_{i=1}^{n} \left(\frac{\partial H}{\partial q_i} \dot{q}_i + \frac{\partial H}{\partial p_i} \dot{p}_i \right).$$

By the virtue of the canonical equations

$$\frac{dH}{dt} = 0,$$

from which it follows that

$$H = \text{constant} = E_{total}.$$

This is the law of conservation of energy, stating that for a conservative system the total energy (which is the Hamiltonian) is constant.

10.3.2   Newton's law of motion

We consider the simplest mechanical system of a single particle, but since any complex mechanical system may be considered a collection of many particles, the following is valid for those as well. Let the mass of the particle be m and its position defined at a certain time t by the displacements:

$$q_i(t), i = 1, 2, 3,$$

where $q_1(t) = x(t), q_2(t) = y(t), q_3(t) = z(t)$.

The kinetic energy of the particle is then

$$E_k = \sum_{i=1}^{3} \frac{1}{2} m \dot{q}_i^2.$$

The particle is moving from its position at time t_0 to a position at time t_1. We assume that the system is conservative, where the change in kinetic energy is equalized by the change in potential energy, and Hamilton's principle applies,

$$\int_{t_0}^{t_1} L dt = \text{extremum}.$$

Applying the Euler-Lagrange equation for each $i = 1, 2, 3$

$$\frac{\partial L}{\partial q_i} - \frac{d}{dt} \frac{\partial L}{\partial \dot{q}_i} = 0.$$

Since

$$L = E_k - E_p.$$

and the kinetic energy is function of the velocities while the potential energy is that of the displacements, the Euler-Lagrange equations become

$$\frac{d}{dt}\frac{\partial E_k}{\partial \dot{q}_i} = -\frac{\partial E_p}{\partial q_i}.$$

For a conservative system, there exists a force potential such that

$$f_i = -\frac{\partial E_p}{\partial q_i}.$$

Here f_i are the components of the force in the coordinate directions. Substituting, executing the differentiation and reordering yields

$$f_i = m\ddot{q}_i; i = 1, 2, 3.$$

This is Newton's second law of motion, better known in the form of

$$F = ma.$$

Let us illustrate this by describing the motion of a mass suspended by a rigid bar from a fixed point and allowing it to rotate in the plane, a pendulum.

Using polar coordinates, the mass at a certain time is at (r, θ), where r is the length of the bar and the position of the pendulum is at the angle θ. The kinetic energy of the mass (assuming the bar is massless) is

$$E_k = \frac{1}{2}mr^2\dot{\theta}^2.$$

The potential energy is simply the height of the mass over the neutral (bottom) position

$$E_p = mgr\cos(\theta).$$

Differentiating the relevant terms

$$\frac{d}{dt}\frac{\partial E_k}{\partial \dot{\theta}} = mr^2\ddot{\theta}$$

and

$$\frac{\partial E_p}{\partial \theta} = mgr\sin(\theta).$$

Substituting and shortening yields the governing equation of motion for the pendulum

$$\ddot{\theta} + \frac{g}{r}\sin(\theta) = 0.$$

This is of course a single degree of freedom system since the only freedom for the pendulum is the rotation about the fixed point on a rigid bar.

10.4 Lagrange's equations of motion

Let us now extend Hamilton's principle by including external work:

$$\int_{t_0}^{t_1} (E_k - (E_p - W_e))\, dt = \text{extremum.}$$

The external work asserted on the system is of the form

$$W_e = q_i f_i^a.$$

Here f_i^a is the force acting on the mass point m in the three coordinate directions. Note that the force components may be time dependent and the work depends on the displacement of the system. Furthermore, let us generalize the potential energy as

$$E_p = \frac{1}{2} k_i q_i^2,$$

where the k_i term represents some way of storing potential energy, for example a spring. The amount of energy stored, similarly to the pendulum in the last section, is related to the position of the mass relative to the stationary position.

Taking the above into consideration, the Euler-Lagrange equation of the extended form becomes

$$-\frac{\partial E_p}{\partial q_i} - \frac{d}{dt}\frac{\partial E_k}{\partial \dot{q}_i} + \frac{\partial W_e}{\partial q_i} = 0,$$

which brings Lagrange's equations of motion

$$\frac{d}{dt}\frac{\partial E_k}{\partial \dot{q}_i} + \frac{\partial E_p}{\partial q_i} = \frac{\partial W_e}{\partial q_i}.$$

Hence, following the last section and executing the derivatives, we obtain

$$m\ddot{q}_i + k_i q_i = f_i^a; i = 1, 2, 3.$$

These equations represent the forced vibrations of the system of a single mass particle. In addition to that, lack of external forces brings the free (natural) vibrations of the system

$$m\ddot{q}_i + k_i q_i = 0; i = 1, 2, 3,$$

crucial in many industrial applications. These will be used in several later sections.

Finally, it is possible to allow dissipative forces resulting in non-conservative systems. The dissipative forces are the negative gradient of the Rayleigh dissipation function as

$$f_i^d = -\frac{\partial D}{\partial \dot{q}_i}.$$

The Rayleigh dissipative function is of the form

$$D = \frac{1}{2} \sum_{i=1}^{3} d_i \dot{q}_i^2,$$

where d_i are the coefficients of dissipation. Hence, the dissipative forces are

$$f_i^d = -d_i \dot{q}_i.$$

Adding these forces to the right-hand side we obtain the extended form of Lagrange's equations of motion (or sometimes called Lagrange's equations of the second kind) as

$$\frac{d}{dt} \frac{\partial E_k}{\partial \dot{q}_i} + \frac{\partial E_p}{\partial q_i} = \frac{\partial W_e}{\partial q_i} - \frac{\partial D}{\partial \dot{q}_i}.$$

Substituting the above derivatives yields

$$m\ddot{q}_i + d_i \dot{q}_i + k_i q_i = f_i^a; i = 1, 2, ..., 3$$

These equations describe the forced, damped vibrations of a single mass particle. Assuming only a single direction, for example x motion, the single mass-spring-damper system is shown in Figure 10.1.

Again, lack of right-hand side forces brings the free, damped vibrations of the system,

$$m\ddot{q}_i + d_i \dot{q}_i + k_i q_i = 0; i = 1, 2, 3,$$

crucial in many industrial applications. These will also be used in several later sections.

10.4.1 Mechanical system modeling

Lagrange's equations of motion are widely used for modeling mechanical systems. The single mass, but three degrees of freedom system developed in the last section may be generalized to a mechanical system of n distinct masses $m^j, j = 1, ..., n$ and the generalized coordinates of the motion become

$$q_1 = x_1, q_2 = y_1, q_3 = z_1; q_4 = x_2, q_5 = y_2, q_6 = z_2; ...$$

and

$$q_{3n-2} = x_n, q_{3n-1} = y_n, q_{3n} = z_n.$$

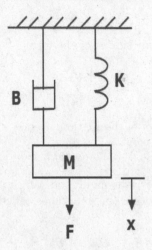

FIGURE 10.1 Mass-spring-damper system

They are gathered into a vector q. The masses of the particles are

$$m_1 = m_2 = m_3 = m^1; m_4 = m_5 = m_6 = m^2; \ldots$$

and

$$m_{3n-2} = m^n, m_{3n-1} = m^n, m_{3n} = m^n.$$

Note the distinction between the subscripts and superscripts. In a general system the particles may be connected via an energy retaining component (for example springs, k_i) and simultaneously be damped by some kind of a dissipative phenomenon (for example, friction, d_i).

For illustration, let us consider a system of two particles by duplicating the system shown in Figure 10.1. The second system is attached to the first mass in place of the force, and we have the force $f(t)$ acting on the second mass only. The masses, springs and dampers may be different between the two systems,

$$m_1 \neq m_2, k_1 \neq k_2, d_1 \neq d_2.$$

The matrix equation for this system is assembled as

$$\begin{bmatrix} m_1 & 0 \\ 0 & m_2 \end{bmatrix} \begin{bmatrix} \ddot{x}_1 \\ \ddot{x}_2 \end{bmatrix} + \begin{bmatrix} d_1 + d_2 & -d_2 \\ -d_2 & d_2 \end{bmatrix} \begin{bmatrix} \dot{x}_1 \\ \dot{x}_2 \end{bmatrix} + \begin{bmatrix} k_1 + k_2 & -k_2 \\ -k_2 & k_2 \end{bmatrix} \begin{bmatrix} x_1 \\ x_2 \end{bmatrix} = \begin{bmatrix} 0 \\ f(t) \end{bmatrix}.$$

With a larger number of mass points and not just a linear but planar or spatial connectivity arrangement, the generalization leads to modeling and analysis of large systems resulting in a matrix equation of the form

$$M\underline{\ddot{q}} + D\underline{\dot{q}} + K\underline{q} = \underline{f}^a.$$

Here M, D, K are the mass, damping and stiffness matrices representing all masses, springs and damping in the system, the $\underline{q}$ is the vector containing all mass point positions and $\underline{f}^a$ contains all active forces. Ultimately this is also the basis for modeling continuum systems discussed in Chapter 12.

10.4.2 Electro-mechanical analogy

Let us consider the behavioral characteristics of mechanical systems and focus on a single mass system for simplicity. We can say that the mass m is representing the inertia, the spring k is the elastic energy and the damping d is the dissipative function. The inertia may be compared to the inductance of a coil in an electrical circuit, denoted by L. The damping is analogous to the resistance R of the circuit, and the elastic spring is akin to the capacitor C of a circuit.

To complete the analogy, we can associate the displacement of the mass in the mechanical system with the charge of the electrical circuit. With these analogies, we can write the single degree of freedom mechanical system governing equation from Section 10.4 in electrical terms as

$$L\frac{d^2Q}{dt^2} + R\frac{dQ}{dt} + \frac{1}{C}Q = E(t).$$

Here the E represents the external power source of the circuit and as such, it is analogous to the active force in the mechanical system. The capacitor is inversely proportional to the spring in the mechanical system. Figure 10.2 shows the electrical circuit corresponding to this equation. The arrow indicates the direction of current, and the open segment between the + and − signs is the location of the external power source.

As a consequence of this analogy, a simple mechanical system may be electronically simulated by creating a circuit replacing the inertia with inductance, damping with resistance, stiffness with the reciprocal of a capacity, and the actuating mechanical force with the electromotive force applied to the circuit. This was the founding principle of the analog computers of the 1950s.

In that process, the electro-mechanical analogy was carried into complex systems with multiple connected circuits to simulate multiple member mechanical system behavior. Hence, Lagrange's equation of motion can also be generalized to electrical systems.

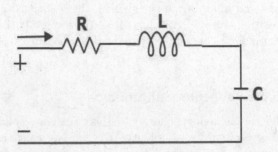

FIGURE 10.2 Electrical circuit

As kinetic energy, we introduce the magnetic energy as

$$E_m = \frac{1}{2} \sum_{i=1}^{n} \sum_{j=1}^{n} L_{ij} \dot{Q}_i \dot{Q}_j,$$

the electrostatic energy as potential energy

$$E_s = \frac{1}{2} \sum_{i=1}^{n} \sum_{j=1}^{n} \frac{1}{C_{ij}} Q_i Q_j,$$

and dissipation function is represented by the energy loss in the resistors as

$$D = \frac{1}{2} \sum_{i=1}^{n} \sum_{j=1}^{n} R_{ij} \dot{Q}_i \dot{Q}_j.$$

Here D is defined as half of the energy transformed into heat per unit time. Hence Lagrange's equation of motion, without an external power source will become

$$\frac{d}{dt} \frac{\partial E_m}{\partial \dot{Q}_i} + \frac{\partial E_s}{\partial Q_i} + \frac{\partial D}{\partial \dot{Q}_i} = 0.$$

The actual current flowing through the circuit is

$$I = \frac{dQ}{dt}.$$

In practice, another form of the circuit equation, in terms of the current, is preferred since that is an easily measurable quantity, but our focus being on the mechanical analogy, we used the charge based form.

10.5 Orbital motion

We now address the orbital motion of two celestial bodies moving under each other's gravitational influence. It is known from Newtonian mechanics that such a motion is planar and we can confine our discussion to the $x - y$ plane.

We will assume that the central body is located at the origin and the moving body has unit mass. The location of the moving body at time t is at coordinates x, y, and it is moving on a path

$$u(x, y, t).$$

The distance between these bodies is

$$\sqrt{x^2(t) + y^2(t)}.$$

Newton's law of gravitation states that the gravitational potential acting on the moving body is

$$E_p = -\frac{\gamma}{\sqrt{x^2 + y^2}}.$$

Here the constant γ is the universal gravitational constant generated by the mass of the central body. The velocity of the orbiting body is

$$\sqrt{(\dot{x})^2 + (\dot{y})^2},$$

hence its kinetic energy is

$$E_k = \frac{1}{2}((\dot{x})^2 + (\dot{y})^2).$$

Let us first observe this in the Lagrangian framework. The Lagrangian becomes

$$L = \frac{1}{2}((\dot{x})^2 + (\dot{y})^2) - \frac{-\gamma}{\sqrt{x^2 + y^2}}.$$

We have two generalized displacement variables and their velocities as

$$q_1 = x, q_2 = y, \dot{q}_1 = \dot{x}, \dot{q}_2 = \dot{y}.$$

One set of components of Lagrange's equations of motion are obtained by evaluating

$$\frac{d}{dt}\frac{\partial L}{\partial \dot{q}_i}; i = 1, 2.$$

The results are $\ddot{x}$ and $\ddot{y}$. The other component set comes from computing

$$\frac{\partial L}{\partial q_i}, i = 1, 2,$$

and produces the expressions

$$-\gamma x(x^2 + y^2)^{-3/2}, -\gamma y(x^2 + y^2)^{-3/2}.$$

Hence, the two Lagrange equations of motion are

$$\ddot{x} + \gamma x(x^2 + y^2)^{-3/2} = 0$$

and

$$\ddot{y} + \gamma y(x^2 + y^2)^{-3/2} = 0.$$

Now turning to Hamilton's formulation, we introduce the variables

$$p_i = \frac{dq_i}{dt}.$$

Specifically in our case they are

$$p_1 = \dot{x}, p_2 = \dot{y}.$$

The other generalized variables remain as $q_1 = x, q_2 = y$. The Hamiltonian is, as shown in an earlier section, of the form

$$H = \frac{1}{2}(p_1^2 + p_2^2) + \frac{-\gamma}{\sqrt{q_1^2 + q_2^2}}.$$

The canonical equations are

$$\dot{p}_i = -\frac{\partial H}{\partial q_i} = -\gamma q_i(q_1^2 + q_2^2)^{-3/2}; i = 1, 2,$$

and

$$\dot{q}_i = \frac{\partial H}{\partial p_i} = p_i; i = 1, 2.$$

The resulting canonical system of four equations becomes

$$p_1 = \frac{dx}{dt},$$

$$p_2 = \frac{dy}{dt},$$

$$\dot{p}_1 = -\gamma x(x^2 + y^2)^{-3/2},$$

and

$$\dot{p}_2 = -\gamma y(x^2 + y^2)^{-3/2}.$$

It is easy to see by differentiation and substitution that the solution of this system is identical to the Lagrange solution.

Polar coordinate transformation enables us to further simplify this problem. The kinetic energy per unit mass is then

$$E_k = \frac{1}{2}(\dot{r}^2 + r^2\dot{\phi}^2)$$

and the gravitational potential (also per unit mass) energy becomes

$$E_p = -\gamma\frac{1}{r}.$$

Let us use this formulation for the solution of the problem. The variational statement using Hamilton's principle becomes

$$\int_{t_1}^{t_2} \left(\frac{1}{2}(\dot{r}^2 + r^2\dot{\theta}^2) + \frac{\gamma}{r} \right) dt = \text{extremum}.$$

The first Euler-Lagrange differential equation for this is

$$\frac{\partial f}{\partial r} - \frac{d}{dt}\frac{\partial f}{\partial \dot{r}} = r\dot{\theta}^2 - \frac{\gamma}{r^2} - \ddot{r} = 0.$$

Reorganizing, we get the ordinary differential equation

$$\frac{d^2r}{dt^2} - r\left(\frac{d\theta}{dt}\right)^2 = -\frac{\gamma}{r^2}.$$

The second Euler-Lagrange differential equation becomes

$$\frac{\partial f}{\partial \theta} - \frac{d}{dt}\frac{\partial f}{\partial \dot{\theta}} = 2r\frac{dr}{dt}\frac{d\theta}{dt} + r^2\frac{d^2\theta}{dt^2} = \frac{d}{dt}\left(r^2\frac{d\theta}{dt}\right) = 0$$

from which it follows that

$$r^2\frac{d\theta}{dt} = c.$$

The value of the constant is the angular momentum per unit mass

$$l = c.$$

Hence, the second differential equation becomes

$$\frac{d\theta}{dt} = \frac{l}{r^2}.$$

Substituting into the first results in

$$\frac{d^2r}{dt^2} - r\left(\frac{l}{r^2}\right)^2 = -\frac{\gamma}{r^2}.$$

Introducing a new variable

$$u = \frac{1}{r},$$

then differentiating brings

$$\frac{dr}{dt} = \frac{d}{dt}\left(\frac{1}{u}\right) = -\frac{1}{u^2}\frac{du}{d\theta}\frac{d\theta}{dt} = -l\frac{du}{d\theta}.$$

Similarly

$$\frac{d^2r}{dt^2} = \frac{d}{d\theta}\frac{dr}{dt}\frac{d\theta}{dt} = \frac{d}{d\theta}\left(-l\frac{du}{d\theta}\right)lu^2 = -l^2u^2\frac{d^2u}{d\theta^2}.$$

Hence the governing equation becomes

$$-l^2u^2\frac{d^2u}{d\theta^2} - \frac{1}{u}l^2u^4 = -\gamma u^2,$$

which after appropriate shortening by common terms simplifies to

$$\frac{d^2}{d\theta^2} + u = \frac{\gamma}{l^2}.$$

This is a non-homogeneous second order ordinary differential equation with solution of

$$u = \frac{\gamma}{l^2} + c_1\cos(\theta) + c_2\sin(\theta).$$

Assuming that the initial position is a maximum or minimum of the orbit

$$u' = -c_1\sin(\theta) + c_2\cos(\theta) = 0$$

implies

$$c_2 = 0.$$

A second initial condition may be posed on the original variable as

$$r(\theta = 0) + r(\theta = \pi) = 2a.$$

Substituting

$$2a = \frac{l^2}{\gamma}\left(\frac{1}{1 + c_3\cos(0)} + \frac{1}{1 + c_3\cos(\pi)}\right) = \frac{l^2}{\gamma}\frac{2}{1 - c_1^2}.$$

Hence

$$c_3 = \sqrt{1 - \frac{l^2}{\gamma a}},$$

which produces

$$r = \frac{l^2}{\gamma}\frac{1}{1 + \sqrt{1 - \frac{l^2}{\gamma a}}\cos(\theta)}.$$

Introducing the energy

$$\epsilon = \frac{-\gamma}{2a}$$

and the eccentricity

$$e = \sqrt{1 + \frac{2\epsilon l^2}{\gamma^2}}$$

the orbit equation becomes that of a conic section

$$r(\theta) = \frac{l^2}{\gamma} \frac{1}{1 + e\cos(\theta)}.$$

This equation describes the path of the orbiting body in relationship to the one in the center.

Specifically the orbital trajectory is an ellipse when $e < 1$, a parabola when $e = 1$, and a hyperbola when $e > 1$. The minimum value of the elliptic trajectory radius is

$$r_{min} = \frac{\theta^2}{\gamma} \frac{1}{1 + e},$$

while its maximum is at

$$r_{max} = \frac{\theta^2}{\gamma} \frac{1}{1 - e}.$$

These are the lengths of the minor and major axes of the elliptical path of the orbiting body.

10.5.1 Conservation of angular momentum

This is in essence the conservation of the energy principle introduced earlier applied to a rotational system.

The angular momentum L of a particle on a circular path is a vector quantity defined as

$$\underline{L} = \underline{r} \times \underline{p}$$

where r is the radius of the circular motion as shown in Figure 10.3. The rotation results in a tangential linear momentum of the particle, defined as

$$\underline{p} = m\underline{v}$$

with $\underline{v}$ being the circular velocity. The rate of change of the angular momentum is

$$\frac{d\underline{L}}{dt} = \frac{d}{dt}(\underline{r} \times \underline{p}).$$

Using the product rule produces

$$\frac{d\underline{L}}{dt} = \frac{d\underline{r}}{dt} \times \underline{p} + \underline{r} \times \frac{d\underline{p}}{dt}.$$

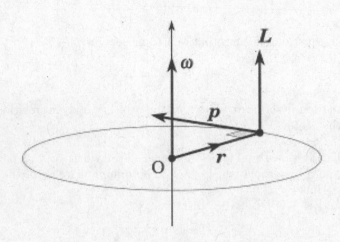

FIGURE 10.3 Angular momentum

Since

$$\frac{d\underline{r}}{dt} = \underline{v},$$

the first term becomes

$$\underline{v} \times \underline{p} = m(\underline{v} \times \underline{v}) = 0.$$

The second term using Newton's law changes as

$$\frac{d\underline{p}}{dt} = m\frac{d\underline{v}}{dt} = m\underline{a} = \underline{F}$$

and also becomes zero,

$$\underline{r} \times \frac{d\underline{p}}{dt} = \underline{r} \times \underline{F} = 0,$$

since $\underline{r}$ and $\underline{F}$ are parallel. Hence

$$\frac{dL}{dt} = 0$$

which implies that the angular momentum is preserved.

10.5.2   The 3-body problem

This is a famous, in general unsolved problem of orbital modeling, subject to the intense efforts by scientists like Lagrange, who spent decades of working on it. The problem consists of 3 celestial bodies with their respective gravitational potential fields and their joint motion.

The reason for the discussion here is that a calculus of variables foundation enables the solution of a so-called restricted version of the problem. This restriction, in part, is focusing on a planar version that enables the finding of extremely important orbital locations, the Lagrange points named after his work and discovery.

In the 3-body problem these objects have different masses. The largest one on the left of Figure 10.4 is assumed to have a mass of M and is located in the origin of the x-axis. The second largest mass (m) is at the distance of R on the right, also on the x-axis. They both have measurable gravitational fields. Finally, the third object is at the location of a certain x value (L_1), and is assumed to be of negligible mass (the second part of the restriction) with no gravitational field, rotating in synchrony with the second mass on its right.

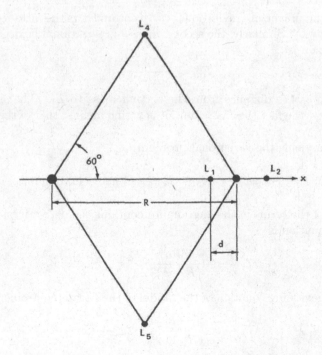

FIGURE 10.4 3-body problem

The cover image is an artistic rendering of the equipotential lines of the 3-body gravitational scenario. The intersection points of the curves are the L_1, L_2, L_3 point locations. The center points of the two interior circular shapes are the location of the large (M) and small (m) masses, respectively, from the right. The two oblique shaped areas are the home of the Lagrange points L_4, L_5.

Let us now introduce a center of mass point that is somewhere between the two masses on the x-axis at the distance of r from the main mass on the left. That is the center of rotation for all masses participating in the system.

With this comes a potential function for the centripetal force in this rotating system acting at the third body position is

$$E_c = -\frac{1}{2}(x - r)^2 \omega^2,$$

where $x - r$ is the third body's distance from the center of rotation.

Following the prior section, the gravitational field potential of the large mass is of the form

$$E_M = \frac{GM}{x},$$

where G is the universal gravitational constant and x is the distance from the large mass to L_1. Similarly, the second mass's gravitational field potential at the L_1 location is

$$E_m = -\frac{Gm}{R - x},$$

where $R - x$ is the distance from the second mass to L_1. The sign is negative to reflect the fact that its potential is acting against that of the big mass.

Their sum yields the variational problem

$$I = \int L dx = \int (E_M + E_m + E_c) dx = \text{extremum}.$$

Since none of the terms in the Lagrangian contains the derivatives, the Euler-Lagrange equation is

$$\frac{\partial L}{\partial x} = \frac{GM}{x^2} - \frac{Gm}{(R - x)^2} - (x - r)\omega^2 = 0.$$

Hence, the governing equation of the model of the restricted problem becomes

$$G\frac{M}{x^2} - G\frac{m}{(R - x)^2} = (x - r)\omega^2.$$

Substituting

$$\omega = \frac{2\pi}{T},$$

dividing by G and substituting Kepler's 3rd law of

$$T^2 = \frac{4\pi^2}{G(M+m)}R^3$$

results in

$$\frac{M}{x^2} - \frac{m}{(R-x)^2} = \frac{(M+m)(x-r)}{R^3}.$$

Dividing by M, introducing

$$k = \frac{m}{M}$$

and using common denominator results in the algebraic equation

$$(R-x)^2 R^3 - kx^2 R^3 - (k+1)x^2(R-x)^2(x-r) = 0.$$

This is an equation containing a parameter k representing the ratio of the participating masses. Hence the locations of the Lagrange points are different between any pairs of celestial bodies. Furthermore, it is of fifth order in the unknown variable x, hence it is unsolvable analytically, as was proven by Ruffini, Abel and Galois centuries ago. For a given k value, however, it is possible to find a numerical solution for the location of the L_1 point.

The distance of the L_1 point from the smaller mass is obtained approximately as

$$R - x \approx R \sqrt[3]{\frac{1}{3\left(\frac{1}{k}+1\right)}}.$$

For illustration, we consider the Sun (M) and Earth (m) scenario. The ratio of their masses is approximately

$$k \approx 3 \cdot 10^{-6},$$

or the Earth's mass is about three millionths of that of the Sun. Substituting this value and executing the arithmetics yield

$$R - x \approx R\sqrt[3]{10^{-6}} = R \cdot 10^{-2} = 0.01R,$$

meaning that the distance of the L_1 point from Earth (denoted by d in Figure 10.4) is about 1 % of the way from the Earth to Sun.

The equation, being of fifth order, produces two more real roots in the line of the masses along our x-axis, they are called the L_2, L_3 points. As the figure shows, L_2 is on the other side of the smaller mass (m), while L_3 is at a distance on the other side of the larger mass (M), hence not shown on the figure.

Finally the complex pair leads to two points located outside of the x-axis at the intersection of the line originating in the location of the large mass, inclining ± 60 degrees from the x-axis and the circular orbit of the smaller (m) mass. These are the L_4, L_5 points.

The practical importance of these locations is in the fact that they are stationary points of the combined gravitational fields. Objects placed in those points will stay synchronized to the motion of the outside (m) mass. Specifically L_4, L_5 are stable; the other ones are saddle points of the combined gravitational fields, hence not fully stable.

Nature has recognized these locations by asteroids captured in Jupiter's L_4, L_5 points, called the Trojan and Greek asteroids. Humankind also exploited these locations: our SOHO solar observatory is located in the L_1 point between Sun and Earth, approximately 1.5 million kilometers from Earth. That is 1 % of the average 150 million kilometers between Earth and Sun. L_2 is also about the same distance from Earth on the opposite side from Sun in Earth's shadow, hence well positioned to observe outside of our Solar system. The new generation Webb space telescope will be located in the L_2 point after its 2021 launch.

10.6　　Variational foundation of fluid motion

Until now we have focused on particles of mechanical systems. To provide a foundation for a later topic, we now consider a fluid "particle" in the form of an infinitesimally small volume ν. For simplicity of the presentation, the vector notation will be omitted since the use of vector operations distinguishes the vector components.

Let us now follow the Hamiltonian avenue again. The mass of the infinitesimal volume of fluid is

$$dm = \rho \, d\nu,$$

where ρ is the density of the fluid. Then, the kinetic energy of the infinitesimal mass of fluid is

$$e_k = \frac{1}{2}\rho|v|^2 \, d\nu.$$

The potential energy of the fluid element is in the form

$$e_p = \rho\phi \, d\nu,$$

where ϕ is the gravitational potential. The lower case e letters indicate the energy of the small fluid volume as opposed to the total fluid. The variational form of our problem then is

$$\int_{t_1}^{t_2} (e_k - e_p)dt = \text{extremum}.$$

Combining all the elementary fluid volumes and substituting yields

$$\int_{t_1}^{t_2} \int_\nu \rho \left(\frac{1}{2}|v|^2 - \phi \right) d\nu dt = 0.$$

We will assume that this elementary volume of fluid will not change but, true to the behavior of fluid, could move by the displacement vector $u = (u_x, u_y, u_z)$. The condition of the unchanged volume may be expressed as

$$\int_\nu \left(\frac{\partial u_x}{\partial x} + \frac{\partial u_y}{\partial y} + \frac{\partial u_z}{\partial z} \right) d\nu = \int_\nu \nabla \cdot u \, d\nu = 0.$$

This condition we apply with a yet unknown Lagrange multiplier λ. Exploiting the identity

$$\nabla \cdot (\lambda u) = \lambda(\nabla \cdot u) + \nabla(\lambda) \cdot u,$$

we obtain

$$\lambda(\nabla \cdot u) = \nabla \cdot (\lambda u) - \nabla(\lambda) \cdot u.$$

Substituting into the volume condition

$$\int_\nu \lambda(\nabla \cdot u)d\nu = \int_\nu (\nabla \cdot (\lambda u) - \nabla(\lambda) \cdot u)d\nu = 0.$$

The first term may be transformed into a surface integral of the volume and as such vanishes, hence the second term represents the constraint in the variational problem

$$\int_{t_1}^{t_2} \int_\nu \left(\rho \left(\frac{1}{2}|v|^2 - \phi \right) - \nabla(\lambda) \cdot u \right) d\nu dt = \text{extremum}.$$

The Euler-Lagrange differential equation corresponding to this constrained variational problem now becomes

$$-\rho \frac{dv}{dt} - \rho\nabla(\phi) - \nabla(\lambda) = 0.$$

By reordering we obtain

$$\frac{dv}{dt} = -\nabla(\phi) - \frac{1}{\rho}\nabla(\lambda).$$

What remains to be found is the physical meaning of the Lagrange multiplier. Let us assume that the fluid is in equilibrium, then $v = 0$. The equation then simplifies to

$$\nabla(\phi) + \frac{1}{\rho}\nabla(\lambda) = 0.$$

In the case of incompressible fluid $\rho = $ constant $= \rho_0$ and may be moved into the differential operator. Hence, the equation may be simplified to

$$\phi + \frac{\lambda}{\rho_0} = \text{constant}.$$

The gravitational potential ϕ at a height (or depth as we will see) is

$$\phi = -gz,$$

hence we obtain

$$\lambda = \rho_0 g(z - z_0).$$

The integration constant above is captured in the reference height z_0. This is really Archimedes' law of hydrostatics, known in the form of

$$p = \rho_0 g(z - z_0).$$

Hence, the physical meaning of the Lagrange multiplier is the pressure p.

If we relinquish the incompressibility condition but assume that the density is a function of the pressure, then

$$\frac{\nabla p}{\rho} = \frac{\nabla p}{f(p)} = \nabla P,$$

where

$$P = \int \frac{dp}{f(p)}.$$

The hydrostatic equilibrium is then

$$\phi + P = \text{constant}.$$

For isothermic (constant temperature) fluids the form

$$P = \frac{p_0}{\rho_0} \log \frac{\rho}{\rho_0}$$

applies, resulting in

$$p = p_0 e^{-\alpha z}.$$

This is the gravitational potential based fluid pressure solution and α is a constant specific to the fluid medium. For air, it is $0.1184 km^{-1}$ resulting in Laplace's atmospheric formula of

$$p = p_0 e^{-0.1184z},$$

where z is measured in kilometers and p_0 is the pressure at sea level. The negative exponent indicates the decrease of atmospheric pressure at higher elevations.

In fluid dynamics applications at the same height the gravity potential is constant and its derivative vanishes. Using this, and introducing the pressure instead of the multiplier in our above solution and further differentiation yields

$$\rho \ddot{u} = -\nabla p,$$

which is the well-known Euler equation of fluid dynamics. This will be the starting equation of the computational formulation discussion in Section 12.4.

11

Analytic mechanics

Analytic mechanics is a mathematical science, but it is of high importance for engineers as it provides analytic solutions to fundamental problems of engineering mechanics. At the same time, it establishes generally applicable procedures. Mathematical physics texts, such as [9], laid the foundation for these analytic approaches addressing physical problems.

This chapter presents modeling applications for classical mechanical problems of elasticity utilizing Hamilton's principle. The most fitting application is the excitation of an elastic system by displacing it from its equilibrium position. In this case, the system will vibrate with a frequency characteristic to its geometry and material, while constantly exchanging kinetic and potential energy. The behavior of strings, membranes and beams will be discussed in detail.

The case of non-conservative systems, where energy loss may occur due to dissipation of the energy, will not be discussed. Hamilton's principle may be extended to non-conservative systems, but the added difficulties do not enhance the discussion of the variational aspects, which is our main focus.

11.1   Elastic string vibrations

We first consider the phenomenon of vibration of an elastic string. Let us assume that the equilibrium position of the string is along the x axis, and the endpoints are located at $x = 0$ and $x = L$. We will stretch the string (since it is elastic) by ΔL resulting in a certain force F exerted on both endpoints to hold it in place. We assume there is no loss of energy and the string will vibrate indefinitely if displaced, i.e., the system is conservative.

The particle of the string located at the coordinate value x at the time t has a yet unknown displacement value of $y(x, t)$. The boundary conditions are:

$$y(0, t) = y(L, t) = 0,$$

in other words, the string is fixed at the ends. In order to use Hamilton's principle, we need to compute the kinetic and potential energies.

With unit length mass of ρ, the kinetic energy is of the form

$$E_k = \frac{1}{2} \int_0^L \rho \left(\frac{\partial y}{\partial t} \right)^2 dx.$$

The potential energy is retained in the elongated (stretched) string. The arc length of the elastic string is

$$\int_0^L \sqrt{1 + \left(\frac{\partial y}{\partial x} \right)^2} \, dx,$$

and the elongation due to the transversal displacement is

$$\Delta L = \int_0^L \sqrt{1 + \left(\frac{\partial y}{\partial x} \right)^2} \, dx - L.$$

Assuming that the elongation is small, i.e.,

$$\left| \frac{\partial y}{\partial x} \right| < 1,$$

it is reasonable to approximate

$$\sqrt{1 + \left(\frac{\partial y}{\partial x} \right)^2} \approx 1 + \frac{1}{2} \left(\frac{\partial y}{\partial x} \right)^2.$$

The elongation by substitution becomes

$$\Delta L \approx \frac{1}{2} \int_0^L \left(\frac{\partial y}{\partial x} \right)^2 dx.$$

Hence, the potential energy contained in the elongated string is

$$E_p = \frac{1}{2} F \Delta L = \frac{F}{2} \int_0^L \left(\frac{\partial y}{\partial x} \right)^2 dx.$$

We are now in the position to apply Hamilton's principle. The variational statement describing the phenomenon becomes

$$I(y) = \int_{t_1}^{t_2} (E_k - E_p) dt =$$

$$\frac{1}{2} \int_{t_1}^{t_2} \int_0^L \left(\rho \left(\frac{\partial y}{\partial t} \right)^2 - F \left(\frac{\partial y}{\partial x} \right)^2 \right) dx dt = \text{extremum}.$$

The Euler-Lagrange differential equation for a function of two independent variables, derived in Section 3.3, is applicable and results in

$$F\frac{\partial^2 y}{\partial x^2} = \rho\frac{\partial^2 y}{\partial t^2}.$$ (11.1)

This is the governing equation of the vibration of the elastic string, also known as the **wave equation**.

The solution of the problem may be solved by the separation approach of d'Alembert. We seek a solution in the form of

$$y(x,t) = a(t)b(x),$$

separating it into time and space dependent components. Then

$$\frac{\partial^2 y}{\partial x^2} = b''(x)a(t)$$

and

$$\frac{\partial^2 y}{\partial t^2} = a''(t)b(x),$$

where

$$b''(x) = \frac{d^2 b}{dx^2},$$

$$a''(t) = \frac{d^2 a}{dt^2}.$$

Substituting into Equation (11.1) yields

$$\frac{b''(x)}{b(x)} = \frac{1}{f^2}\frac{a''(t)}{a(t)},$$

where for convenience we introduced

$$f^2 = \frac{F}{\rho}.$$

The two sides of this differential equation are dependent on x and t, respectively. Their equality is required at any x and t values which implies that the two sides are constant. Let us denote the constant by $-\lambda$ and separate the (partial) differential equation into two ordinary differential equations:

$$\frac{d^2 b}{dx^2} + \lambda b(x) = 0,$$

and

$$\frac{d^2 a}{dt^2} + f^2\lambda a(t) = 0.$$

The solution of these equations may be obtained by the techniques learned in Section 5.3 for the eigenvalue problems. The first equation has the spatial solutions of the form

$$b_k(x) = \sin\left(\frac{k\pi}{L}x\right); k = 1, 2, \ldots,$$

corresponding to the eigenvalues

$$\lambda_k = \frac{k^2\pi^2}{L^2}.$$

Applying these values, we obtain the temporal solution from the second equation by means of classical calculus in the form of

$$a_k(t) = c_k \cos\left(\frac{k\pi f}{L}t\right) + d_k \sin\left(\frac{k\pi f}{L}t\right),$$

with c_k, d_k arbitrary coefficients. Considering that at $t = 0$ the string is in a static equilibrium position

$$a'(t = 0) = 0$$

we obtain $d_k = 0$ and the temporal solution of

$$a_k(t) = c_k \cos\left(\frac{k\pi f}{L}t\right).$$

The general mathematical model for the vibrating string becomes

$$y_k(x, t) = c_k \cos\left(\frac{k\pi f}{L}t\right) \sin\left(\frac{k\pi}{L}x\right); k = 1, 2, \ldots.$$

For any specific value of k, the natural frequencies of the string are

$$\omega_k = \sqrt{\lambda_k} = \frac{k\pi}{L}$$

and the corresponding spatial solutions are the natural vibration shapes, also called normal modes:

$$b_k(x) = \sin(\omega_k x).$$

The first three normal modes are shown in Figure 11.1 for an elastic string of unit tension force, mass density, and span. The figure demonstrates that the period of the vibration decreases and the frequency increases for the higher mode number k.

The motion is initiated by displacing the string and releasing it. Let us define this initial enforced amplitude as

$$y(x_m, 0) = y_m,$$

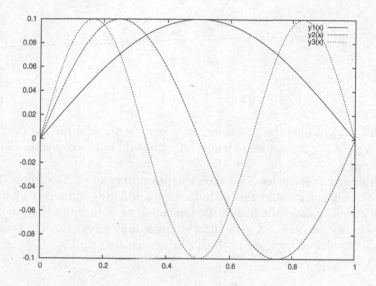

FIGURE 11.1 Normal modes of elastic string

where the x_m describes the location of the initial stationary displacement of the string as an internal value of the span

$$x_m \in (0, L).$$

Then the initial shape of the string is a triangle over the span, described by the function

$$f(x) = \begin{cases} \frac{y_m}{x_m} x; \ 0 \leq x \leq x_m, \\ y_m + \frac{y_m}{x_m - L}(x - x_m); \ x_m < x \leq L. \end{cases}$$

The unknown coefficient may be solved from the initial condition as

$$y(x_m, 0) = f(x_m) = c_k \cos\left(\frac{k\pi f}{L} 0\right) \sin\left(\frac{k\pi}{L} x_m\right) = y_m,$$

from which

$$c_k = \frac{y_m}{\sin\left(\frac{k\pi}{L} x_m\right)}.$$

Note that if the interior point is the middle point of the span,

$$x_m = \frac{L}{2},$$

then the first $(k = 1)$ coefficient will be simply the y_m amplitude:

$$c_1 = y_m,$$

since

$$\sin\left(\frac{\pi}{L}\frac{L}{2}\right) = \sin\left(\frac{\pi}{2}\right) = 1.$$

Similar, but not identical, considerations may be applied for the coefficients of the higher normal modes of which only the odd numbered will exist.

The natural frequencies depend on the physical conditions, such as the pre-applied tension force and the material characteristics embodied in the unit weight ρ. Specifically, the higher the tension force F in the string, the higher the frequency becomes. A very tight string vibrates very quickly (with high frequency), while a very loose string vibrates slowly.

11.2   The elastic membrane

We now turn to the phenomenon of the vibration of an elastic membrane. We assume that the membrane is fixed on its perimeter L which surrounds the domain D of the membrane. We further assume that the initial, equilibrium position of the membrane is coplanar with the $x - y$ plane.

$$z(x, y, t) = 0; t = 0.$$

The membrane is displaced by a certain amount and released. The ensuing vibrations are the subject of our interest. The vibrations are a function of the location of the membrane and the time as

$$z = z(x, y, t).$$

We will again use Hamilton's principle after the kinetic and potential energies of the membrane are found. Let us assume that the unit area mass of the membrane does not change with time, and is not a function of the location:

$$\rho(x, y) = \rho = \text{constant}.$$

The velocity of the membrane point at (x, y) is

$$v = \frac{\partial z}{\partial t},$$

resulting in kinetic energy of

$$E_k = \frac{1}{2} \int \int_D \rho v^2 \, dx dy$$

or

$$E_k = \frac{1}{2} \int \int_D \rho \left(\frac{\partial z}{\partial t}\right)^2 dx dy.$$

We consider the source of the potential energy to be the stretching of the surface of the membrane. The initial surface is

$$\int \int_D dx dy,$$

and the extended surface is

$$\int \int_D \sqrt{1 + \left(\frac{\partial z}{\partial x}\right)^2 + \left(\frac{\partial z}{\partial y}\right)^2} \, dx dy.$$

Assuming small vibrations, we approximate as earlier in the case of the string

$$\sqrt{1 + \left(\frac{\partial z}{\partial x}\right)^2 + \left(\frac{\partial z}{\partial y}\right)^2} \approx 1 + \frac{1}{2}\left(\left(\frac{\partial z}{\partial x}\right)^2 + \left(\frac{\partial z}{\partial y}\right)^2\right).$$

Hence the surface change is

$$\frac{1}{2} \int \int_D \left(\frac{\partial z}{\partial x}\right)^2 + \left(\frac{\partial z}{\partial y}\right)^2 dx dy.$$

The stretching of the surface results in a surface tension σ per unit surface area. The potential energy is the product

$$E_p = \sigma \frac{1}{2} \int \int_D \left(\frac{\partial z}{\partial x}\right)^2 + \left(\frac{\partial z}{\partial y}\right)^2 dx dy.$$

We are now in the position to apply Hamilton's principle. Since

$$I(z) = \int_{t_1}^{t_2} (E_k - E_p)dt = \text{extremum},$$

substitution yields the variational problem of the elastic membrane:

$$\frac{1}{2} \int_{t_1}^{t_2} \int \int_D \left(\rho \left(\frac{\partial z}{\partial t}\right)^2 - \sigma \left(\left(\frac{\partial z}{\partial x}\right)^2 + \left(\frac{\partial z}{\partial y}\right)^2\right)\right) dx dy dt = \text{extremum}.$$

The Euler-Lagrange differential equation for this class of problems following Section 3.5 becomes

$$\sigma \left(\frac{\partial^2 z}{\partial x^2} + \frac{\partial^2 z}{\partial y^2}\right) = \rho \frac{\partial^2 z}{\partial t^2},$$

or using Laplace's symbol

$$\sigma \Delta z = \rho \frac{\partial^2 z}{\partial t^2}.$$

The solution will follow the insight gained at the discussion of the elastic string and we seek a solution in the form of

$$z(x, y, t) = a(t)b(x, y).$$

The derivatives of this solution are

$$\Delta z(x, y, t) = a(t)\Delta b(x, y),$$

and

$$\frac{\partial^2 z(x, y, t)}{\partial t^2} = b(x, y)\frac{d^2 a(t)}{dt^2}.$$

Substitution and separation of terms yields

$$\frac{\sigma \Delta b}{\rho b} = \frac{1}{a(t)}\frac{d^2 a(t)}{dt^2}.$$

Again, since the left-hand side is only a function of spatial coordinates and the right-hand side is only of time, they must be equal and constant, assumed to be $-\lambda$. This separates the partial differential equation into an ordinary differential equation in time,

$$\frac{d^2 a(t)}{dt^2} + \lambda a(t) = 0,$$

and a simpler partial differential equation in (x, y),

$$\sigma \Delta b(x, y) + \lambda \rho b(x, y) = 0.$$

The solution of the first differential equation is

$$a(t) = c_1 \cos(\sqrt{\lambda}t) + c_2 \sin(\sqrt{\lambda}t).$$

Since initially the membrane is in equilibrium,

$$\left.\frac{da}{dt}\right|_{t=0} = 0,$$

which indicates that

$$c_2 = 0.$$

Hence

$$a(t) = c_1 \cos(\sqrt{\lambda}t).$$

In order to demonstrate the solution for the second equation, let us omit the tension and material density for ease of discussion. The differential equation of the form

$$\Delta b(x, y) + \lambda b(x, y) = 0,$$

is the same we solved analytically in the case of the elastic string; however, it is now with a solution function of two variables. The solution strategy will consider the variational form of this eigenvalue problem introduced in Section 5.2:

$$I(b) = \int \int_D \left(\left(\frac{\partial b}{\partial x} \right)^2 + \left(\frac{\partial b}{\partial y} \right)^2 - \lambda b^2(x, y) \right) dxdy = \text{extremum}.$$

11.2.1 Circular membrane vibrations

Let us restrict ourselves to the domain of the unit circle for simplicity. The domain D in rectangular coordinates is defined by

$$D : (1 - x^2 - y^2 \geq 0).$$

We use Kantorovich's method and seek an approximate solution in the form of

$$b(x, y) = \alpha \omega(x, y) = \alpha(x^2 + y^2 - 1),$$

where α is a yet unknown constant. It follows that on the boundary ∂D

$$\omega(x, y) = x^2 + y^2 - 1 = 0,$$

hence the approximate solution satisfies the zero boundary condition. With this choice

$$I(\alpha) = \alpha^2 \int \int_D (4x^2 + 4y^2 - \lambda(x^2 + y^2 - 1)^2) dxdy = \text{extremum}.$$

Introducing polar coordinates for ease of integration yields

$$I(\alpha) = \alpha^2 \int_0^{2\pi} \int_0^1 4r^3 - \lambda r(r^2 - 1)^2 drd\phi = \text{extremum}.$$

The evaluation of the integral results in the form

$$I(\alpha) = (2\pi - \lambda \frac{\pi}{3}) \alpha^2 = \text{extremum}.$$

The necessary condition of the extremum is

$$\frac{\partial I(\alpha)}{\partial \alpha} = 0,$$

which yields an equation for λ

$$2\alpha(2\pi - \lambda \frac{\pi}{3}) = 0.$$

The eigenvalue as the solution of this equation is

$$\lambda = 6.$$

The unknown solution function coefficient may be solved by normalizing the eigensolution as

$$\int\int_D b^2(x,y)dx = 1.$$

Substituting yields

$$\alpha^2 \int_0^{2\pi} \int_0^1 r(r^2 - 1)^2 dr d\phi = 1.$$

Integrating results in

$$\alpha^2 \frac{\pi}{3} = 1.$$

Hence

$$\alpha = \sqrt{\frac{3}{\pi}}.$$

The spatial solution is therefore

$$b(x,y) = \sqrt{\frac{3}{\pi}}(x^2 + y^2 - 1).$$

The complete solution of the differential equation of the elastic membrane of the unit circle is finally

$$z(x,y,t) = c_1 \cos(\sqrt{6}t)\sqrt{\frac{3}{\pi}}(x^2 + y^2 - 1).$$

The remaining coefficient may be established by the initial condition.

Assuming the center of the membrane is displaced by an amplitude A,

$$z(0,0,0) = A = c_1 \sqrt{\frac{3}{\pi}}(-1).$$

from which follows

$$c_1 = -A\sqrt{\frac{\pi}{3}}.$$

The final solution is

$$z(x,y,t) = -A\cos(\sqrt{6}t)(x^2 + y^2 - 1).$$

The shape of the solution is shown in Figure 11.2. The figure shows the solution of the half-membrane at three distinct time steps. The jagged edges are artifacts of the discretization; the shape of membrane was the unit circle.

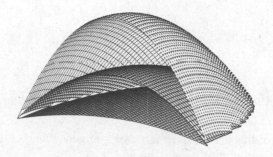

FIGURE 11.2 Vibration of elastic membrane

11.2.2   Non-zero boundary conditions

So far, we restricted ourselves to trivial boundary conditions for the sake of clarity. In engineering practice, however, non-zero boundary conditions are very often imposed. These, also called enforced motion, boundary conditions are the subject of our focus here.

Let us consider the membrane with flexible boundary allowing some or all of the boundary points to attain non-zero displacement from the plane. We introduce $p(s)$ as the tension force in a unit arc length section of the boundary stretched due to a unit displacement of the membrane: $z = 1$. Let the arc length of a section of the boundary in equilibrium be ds. Then, the tension force in the section due to a non-unit displacement z is

$$-p(s)z(x, y, t)ds,$$

where the negative sign indicates the force's effort to pull the boundary back toward the equilibrium position and opposite from the displacement. The potential energy of the boundary section may be computed by

$$p(s)ds \int zdz = \frac{1}{2}p(s)z^2ds.$$

The total potential energy due to the tension force on the boundary L is

$$E_p^L = \frac{1}{2} \int_L p(s)z^2 ds.$$

Applying Hamilton's principle for this scenario now yields

$$I(z) = \frac{1}{2} \int_{t_1}^{t_2} F dt = \text{extremum},$$

where

$$F = \left(\int\int_D \left(\rho \left(\frac{\partial z}{\partial t} \right)^2 - \sigma \left(\left(\frac{\partial z}{\partial x} \right)^2 + \left(\frac{\partial z}{\partial y} \right)^2 \right) \right) dxdy - \int_L p(s)z^2 ds \right).$$

The newly introduced boundary integral's inconvenience may be avoided as follows. First, it may also be written as

$$\int_L p(s)z^2 ds = \frac{1}{2} \int_L \left(p(s)z^2 \frac{ds}{dy} dy + p(s)z^2 \frac{ds}{dx} dx \right).$$

Introducing the twice differentiable

$$P = \frac{1}{2} pz^2 \frac{ds}{dy}$$

and

$$Q = -\frac{1}{2} pz^2 \frac{ds}{dx}$$

functions that are defined on the boundary curve L the integral further changes to

$$\int_L pz^2 ds = \int_L (P dy - Q dx).$$

Finally, with the help of Green's theorem, we obtain

$$\int_L pz^2 ds = \int\int_D \left(\frac{\partial P}{\partial x} + \frac{\partial Q}{\partial y} \right) dxdy.$$

Hence the variational form of this problem becomes

$$I(z) = \frac{1}{2} \int_{t_1}^{t_2} G dt = \text{extremum},$$

where

$$G = \int\int_D \left(\rho \left(\frac{\partial z}{\partial t} \right)^2 - \sigma \left(\left(\frac{\partial z}{\partial x} \right)^2 + \left(\frac{\partial z}{\partial y} \right)^2 \right) - \left(\frac{\partial P}{\partial x} + \frac{\partial Q}{\partial y} \right) \right) dxdy.$$

This problem is identical to the one in Section 3.5, the case of a functional with three independent variables. The two spatial independent variables are augmented in this case with time as the third independent variable. The corresponding Euler-Lagrange differential equation becomes the same as in the case of the fixed boundary

$$\sigma \Delta z = \rho \left(\frac{\partial^2 z}{\partial^2 t} \right),$$

with the addition of the constraint on the boundary as

$$\sigma \frac{\partial z}{\partial n} + pz = 0,$$

where n is the normal of the boundary. The solution may again be sought in the form of

$$z(x, y, t) = a(t)b(x, y),$$

and as before, based on the same reasoning

$$a(t) = c_1 \cos(\sqrt{\lambda}t) + c_2 \sin(\sqrt{\lambda}t).$$

The $b(x, y)$ now must satisfy the following two equations.

$$\sigma \Delta b + \lambda \rho b = 0; (x, y) \in D,$$

and

$$\sigma \frac{\partial b}{\partial n} + pb = 0; (x, y) \in L.$$

The solution of these two equations follows the procedure established in the last section.

11.3 Bending of a beam under its own weight

The two analytic elasticity examples presented so far were one- and two-dimensional, respectively. The additional dimensions (the string's cross-section or the thickness of the membrane) were negligible and ignored in the presentation. In this section we address the phenomenon of the bending of a beam with a non-negligible cross-section and consider all three dimensions.

In order to deal with the problem of the beam, we introduce some basic concepts of elasticity for this specific case only. A fuller exposition of the topic will be in the next chapter. Let us consider an elastic beam with length L and cross-section area A. We consider the beam fully constrained at one end and free on the other, known as a cantilever beam, with a rectangular cross-section of width $2a$ along the z-axis and height $2b$ along the y-axis as shown in Figure 11.3. The axis of the beam is aligned along the x-axis.

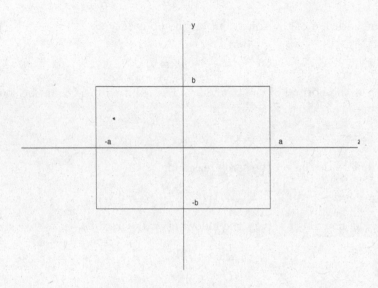

FIGURE 11.3 Beam cross-section

The relationship between the stress resulting from an axial force exerted on the free end of the beam and its subsequent deformation is expressed by the well-known Hooke's law

$$\sigma = E\epsilon,$$

where the constant E, called Young's modulus, expresses the inherent elasticity of the material with regards to elongation. The relationship between the stress (σ) and the force (F) is

$$\sigma = \frac{F}{A}.$$

The strain (ϵ) describes the relative deformation of the beam and in the axial case this is

$$\epsilon = \frac{dl}{l},$$

where dl is the elongation along the beam's longitudinal direction. In different deformation scenarios, like the ensuing bending, the formulation for the strain may vary and will be discussed in more detail later.

The variational statement of this problem is again based on Hamilton's principle; however, since in this particular example we consider a static equilibrium, there is no kinetic energy. The two components of the extended Hamiltonian are the potential energy (manifested in the strain energy) and the work of forces acting on the body.

The internal energy related to the strain is

$$E_s = \frac{1}{2} \int_V \sigma \epsilon dV.$$

Substitution of Hooke's law yields

$$E_s = \frac{1}{2} E \int_V \epsilon^2 dA \, dx.$$

The strain energy of a particular cross-section is obtained by integrating as

$$E_s(x) = \frac{1}{2} E \int_{-b}^{b} \int_{-a}^{a} \epsilon^2 dz dy.$$

The bending will result in a curved shape with a radius of curvature r and a strain in the beam. Note that the radius of curvature is a function of the lengthwise location since the shape of the beam (the subject of our interest) is not a circle.

The relationship between the radius of curvature and the strain is established as follows. Above the neutral plane of the bending, that is the $x - z$ plane in our case, the beam is elongated and it is compressed below the plane. Based on that at a certain distance y above or below the plane the strain is

$$\epsilon = \frac{y}{r}.$$

Note that since y is a signed quantity, above yields zero strain in the neutral plane, positive (tension) above the plane and negative (compression) below. Using this in the strain energy of a particular cross-section yields

$$E_s(x) = \frac{E}{2} \int_{-b}^{b} \int_{-a}^{a} \frac{y^2}{r^2} dz dy = \frac{E}{2} \frac{4ab^3}{3} \frac{1}{r^2} = \frac{EI}{2} \frac{1}{r^2},$$

where

$$I = \frac{4ab^3}{3}$$

is the moment of inertia of the rectangular cross-section with respect to the z-axis. The total strain energy in the volume is

$$E_s = \frac{1}{2}EI \int_0^L \frac{1}{r^2} dx.$$

We assume that the only load on the beam is its weight. We denote the weight of the unit length with w. The bending moment generated by the weight of a cross-section with respect to the neutral, the x-z plane is

$$dM = ywdx,$$

where y is the distance from the neutral plane and dx represents an infinitesimally thin cross-section. The total work of bending will be obtained by integrating along the length of the beam:

$$W = \int_0^L dM = w \int_0^L ydx,$$

since the weight (or density) of the material is constant.

Using Hamilton's principle extended with external work but without a kinetic energy component yields the variational statement of the form

$$I(y) = \int_{t_1}^{t_2} (E_s - \hat{W}) \, dt = \text{extremum}.$$

Substituting the energies brings

$$I(y) = \int_{t_1}^{t_2} \int_0^L \left(\frac{1}{2}EIy''^2(x) - wy \right) dxdt = \text{extremum}.$$

Here we substituted

$$r = \frac{1}{y''(x)}$$

since the radius of curvature is reciprocal of the second derivative of the bent curve of the beam.

Here the functional contains the second derivative; therefore, the Euler-Poisson equation of order two will apply as

$$\frac{\partial f}{\partial y} - \frac{d}{dx}\frac{\partial f}{\partial y'} + \frac{d^2}{dx^2}\frac{\partial f}{\partial y''} = 0.$$

Since in this case

$$f(y, y'') = \frac{1}{2}EIy''^2 - wy,$$

the first term is simply

$$\frac{\partial f}{\partial y} = -w.$$

The second term vanishes as the first derivative of the unknown function is not explicitly present. With

$$\frac{\partial f}{\partial y''} = 2\frac{1}{2}EIy'',$$

the third term becomes

$$\frac{d^2}{dx^2}\frac{\partial f}{\partial y''} = EI\frac{d^4}{dx^4}y.$$

Hence, the governing equation of the phenomenon becomes

$$\frac{d^4y}{dx^4} = \frac{w}{EI}.$$

Direct integration yields the general mathematical model as

$$y(x) = \frac{w}{24EI}(x^4 + 4c_1x^3 + 12c_2x^2 + 24c_3x + c_4),$$

where the c_i are constants of integrations. The solution curve yields the shape of the bent beam shown in Figure 11.4.

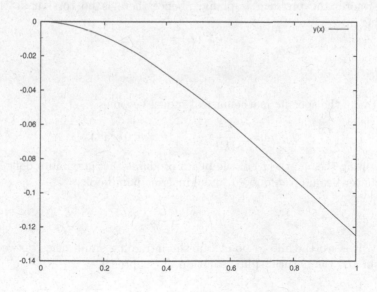

FIGURE 11.4 Beam profile under its weight

In the figure unit physical coefficients were used for the sake of simplicity and the coefficients of integration are resolved from the boundary conditions as follows. At the fixed end, the beam is not deflected, hence

$$y(x = 0) = 0,$$

which implies

$$c_4 = 0.$$

Furthermore, at the fixed end the tangent of the curve is horizontal as

$$y'(x = 0) = 0,$$

implying

$$c_3 = 0.$$

Finally, at the free end the beam has no curvature, the second derivative vanishes. Therefore,

$$y''(L) = 0$$

results in

$$c_2 = \frac{L^2}{2}.$$

Furthermore, the problem is planar, hence there is no torsion at the end resulting in

$$y'''(L) = 0$$

yields

$$c_1 = -L.$$

With these, the specific mathematical model becomes

$$y(x) = \frac{w}{24EI}(x^4 - 4Lx^3 + 6L^2x^2).$$

Substituting the length (L) of the beam produces the' maximum deflection of a cantilever beam often quoted in engineering handbooks:

$$y(L) = \frac{wL^4}{8EI}.$$

Finally, it is worthwhile to point out the intriguing similarities between this problem and the natural spline solution of Chapter 9.

The scenario is also often presented as a problem of optimization. In engineering practice, it is a natural desire to minimize the deflection of the beam under its own weight, since very likely there is an additional load applied to it as well.

In our case, it is easy to see that to minimize the deflection, either of the quantities in the denominator, the Young's modulus or the moment of inertia, should be increased. Assuming that the material type is dictated hence E is fixed, we can still address the shape. Since the chosen cross-section is rectangular, it follows that the higher the b dimension the smaller the deflection is.

This, however, cannot be carried to the extreme, some minimal and maximal ratios of dimensions of rectangular cross-sections are usually given as constraints. The mathematical problem becomes

$$I = \frac{4ab^3}{3} = \text{maximum},$$

subject to

$$r_{min} \leq \frac{b}{a} \leq r_{max}.$$

This is a constrained optimization problem whose solution is intuitively at the maximum ratio

$$I_{max} = \frac{4r_{max}^3 a^4}{3}.$$

This is a simplest problem of shape optimization, a topic of high importance in structural engineering. The method of gradients in Section 6.5 provides the foundation for the variational solution of similar problems.

11.3.1 Transverse vibration of beam

The phenomenon of our interest here is dynamic as opposed to the static nature of the last section. The bending beam was in stationary equilibrium, but now we consider the time dependent behavior of the beam and the solution becomes

$$y = y(x, t).$$

According to our modeling approach of the past sections in this chapter, we establish the potential energy in the elastic beam as

$$E_p = \frac{1}{2} \int_0^L EI \left(\frac{\partial^2 y}{\partial x^2} \right)^2 dx.$$

The kinetic energy of the beam is computed by integrating the kinetic energies of the infinitesimally thin cross-sections of the beam

$$E_k = \frac{1}{2} \int_0^L \rho A \left(\frac{\partial y}{\partial t} \right)^2 dx.$$

Here ρ is the density of the material of the beam and A is the cross-section that we will assume to be constant and unit for the simplicity of the discussion.

Hamilton's principle applied to this scenario results in the corresponding Euler-Lagrange differential equation of

$$\rho \frac{\partial^2 y}{\partial t^2} - \frac{d^2}{dx^2} EI \frac{\partial^2 y}{\partial x^2} = 0.$$

Executing the posted differentiation brings

$$\frac{\partial^2 y}{\partial t^2} - \frac{EI}{\rho} \frac{\partial^4 y}{\partial x^4} = 0,$$

and the governing equation becomes

$$\frac{\partial^2 y}{\partial t^2} = \frac{EI}{\rho} \frac{\partial^4 y}{\partial x^4}.$$

Following d'Alembert's solution approach and the prior sections, we separate

$$y(x,t) = v(t)w(x).$$

Considering the initial condition and following Section 11.1, the temporal solution will be of the form

$$v(t) = \sin(\omega t).$$

For convenience, introduce

$$\gamma^4 = \frac{\rho \omega^2}{EI}.$$

The spatial solution is then obtained from the equation

$$\frac{d^4 w(x)}{dx^4} - \gamma^4 w(x) = 0.$$

The characteristic equation produces the solution

$$w(x) = c_1 e^{\gamma x} + c_2 e^{-\gamma x} + c_3 e^{i\gamma x} + c_4 e^{-i\gamma x}.$$

Using Euler's identities, the more convenient solution form is

$$w(x) = a\cos(\gamma x) + b\sin(\gamma x) + c\cosh(\gamma x) + d\sinh(\gamma x).$$

The coefficients may be resolved by boundary conditions. The beam is fixed on the left-hand side, hence

$$w(0) = 0, w'(0) = 0.$$

Applying these boundary conditions, it follows that

$$a + c = b + d = 0,$$

hence
$$w = a\left(\cosh(\gamma x) - \cos(\gamma x)\right) + b\left(\sinh(\gamma x) - \sin(\gamma x)\right).$$

At the free end, there is no curvature and due to the planar model there is no torsion; therefore,
$$w''(L) = 0, w'''(L) = 0.$$

Applying these conditions will result in the system
$$a\gamma^2\left(\cosh(\gamma L) + \cos(\gamma L)\right) + b\gamma^2\left(\sinh(\gamma L) + \sin(\gamma L)\right) = 0,$$

and
$$a\gamma^2\left(\sinh(\gamma L) - \sin(\gamma L)\right) + b\gamma^2\left(\cosh(\gamma L) + \cos(\gamma L)\right) = 0.$$

The system may be solved if the determinant is zero,
$$\cosh(\gamma L)\cos(\gamma L) + 1 = 0.$$

The solutions of this transcendental equation may be approximated by the values
$$\gamma L = (k - \frac{1}{2})\pi; k = 1, 2, 3, \ldots$$

hence we set
$$\gamma_k = (k - \frac{1}{2})\frac{\pi}{L}; k = 1, 2, 3, \ldots$$

With some algebraic tedium, the vibration shapes corresponding to these values will be of the form
$$w_k = \frac{\cosh(\gamma_k x) - \cos(\gamma_k x)}{\cosh(\gamma_k L) + \cos(\gamma_k L)} - \frac{\sinh(\gamma_k x) - \sin(\gamma_k x)}{\sinh(\gamma_k L) + \sin(\gamma_k L)}.$$

The natural frequencies of vibration are recovered from the earlier definition of γ as
$$\omega_k = \sqrt{\frac{\gamma_k^4 EI}{\rho}} = \gamma_k^2\sqrt{\frac{EI}{\rho}} = ((k - \frac{1}{2})\frac{\pi}{L})^2\sqrt{\frac{EI}{\rho}}; k = 1, 2, 3, \ldots$$

Figure 11.5 shows the first three transversal vibration shapes of the cantilever beam (fixed on one side only) normalized to unit amplitude. The first shape is a single wave going down to the right. The second shape is going to the upper corner after one downward wave. Finally, the third shape exhibits both a lower and an upper wave component.

The mathematical model of the transversal vibration of a beam then becomes
$$y_k(x, t) = c_k w_k(x)\sin(\omega_k t), k = 1, 2, \ldots$$

It is easy to recognize that these vibration shapes, while conceptually sinusoid, adhere to the specific boundary conditions of the problem. The fixed end visibly shows the zero tangent of the waves and at the free end the curves have no curvature as was dictated by the boundary conditions.

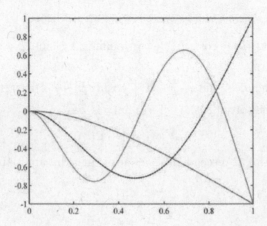

FIGURE 11.5 Transversal vibrations of cantilever beam

11.4   Buckling of a beam under axial load

Another interesting phenomenon related to the flexible beam is its axial behavior that is of utmost importance in the construction industry. The physical problem is to find the limit of the axial loading of a beam without physical failure. The scenario may be described in a horizontal beam with axial loads, but we address the more familiar (albeit mathematically identical) vertical beam buckling case shown in Figure 11.6.

We are going to rely on some of the foundation established in the prior section, specifically we start from the internal strain energy which, for a bent beam, was found to be

$$E_s = \frac{1}{2} \int_0^L EI(y'')^2 dx,$$

where L is the length of the beam, I is the cross-section moment of inertia and E is the Young's modulus of the material. We again consider this as potential energy since if a beam buckled under some load but did not break, the stored strain energy would push it back into straight form after the load is released.

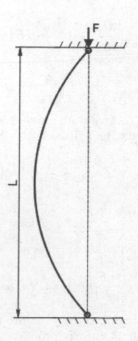

FIGURE 11.6 Buckling scenario

Similarly to the transversal case, we need to consider the work extended on the beam to reach the curved shape, or to generate the potential energy. The axial elongation of the beam is

$$\Delta L \approx \frac{1}{2} \int_0^L y'^2 dx.$$

The work of the axial force F is

$$W_a = F \cdot \Delta L,$$

Hamilton's principle extended to include external work is used here, but in this case lacking kinetic energy:

$$I = \int_{t_1}^{t_2} (E_s - W_a) dt = \text{extremum}.$$

The functional contains a second order derivative, hence the Euler-Poisson equation introduced in Chapter 4 applies, whose components are

$$\frac{d}{dx} \frac{\partial f}{\partial y'} = -\frac{d}{dx}(Fy'),$$

and

$$\frac{d^2}{dx^2}\frac{\partial f}{\partial y''} = \frac{d^2}{dx^2}(EIy''),$$

resulting in

$$-\frac{d}{dx}(Fy') - \frac{d^2}{dx^2}(EIy'') = 0.$$

Assuming that the beam is of constant cross-section and homogenous material, the governing equation may be written as

$$\frac{d^2}{dx^2}y'' + \frac{F}{EI}y'' = 0.$$

Introducing

$$v(x) = y''(x),$$

the buckling differential equation appears as

$$\frac{d^2}{dx^2}v(x) + \frac{F}{EI}v(x) = v''(x) + \frac{F}{EI}v(x) = 0.$$

This is physically meaningful, since y'' is proportional to the inverse of the radius of curvature, that in turn is proportional to the displacement from the axis of the undeformed beam. The solution of such ordinary differential equation by the characteristic equation method, after applying Euler formulae to convert the exponential expressions to trigonometric, is of the form

$$v(x) = A\sin\left(\sqrt{\frac{F}{EI}}x\right) + B\cos\left(\sqrt{\frac{F}{EI}}x\right).$$

Applying the boundary conditions at the ends,

$$v(0) = 0, v(L) = 0$$

means that we do not allow curvature at either end. The boundary condition at the bottom end brings

$$v(0) = 0 \rightarrow B = 0.$$

The top end results in

$$v(L) = A\sin\left(\sqrt{\frac{F}{EI}}L\right) = 0.$$

The non-trivial solution, $A \neq 0$, implies

$$\sin\left(\sqrt{\frac{F}{EI}}L\right) \neq 0,$$

which occurs when

$$\sqrt{\frac{F}{EI}}L = k \cdot \pi; k = 1, 2, ...,$$

The very first such solution is when $k = 1$ and is called the critical load under which the beam first buckles:

$$F_{cr} = \frac{\pi^2 EI}{L^2}.$$

The position attained by the beam when loaded by F_{cr} is stable, if the load will not increase, the shape will not change. The corresponding critical stress value is computed as

$$\sigma_{cr} = \frac{F_{cr}}{A},$$

where A is the cross-section area. If the critical stress value does not exceed the yield point of the material, the deformation is flexible and releasing the load will allow the beam to spring back to its undeformed shape.

What is hidden in our solution is that the actual value of A has not been determined. We only computed the shape of the buckled beam but the midpoint deflection cannot be identified. In order to do so, more physical information must be taken into consideration, and that is beyond our focus here.

11.4.1 Axial vibration of a beam

Let us now turn to the dynamic phenomenon of the axial vibration of the beam. The elastic potential energy in this case is contained in its axial strain,

$$\epsilon = \frac{\partial u}{\partial x},$$

and is proportional to the modulus of elasticity E and cross-section A:

$$E_p = \frac{1}{2} \int_0^L EA \left(\frac{\partial u}{\partial x}\right)^2 dx.$$

Here the displacement along the longitudinal axis is denoted by $u(x, t)$ since the beam is oriented horizontally, along the x axis. The kinetic energy is based on the speed of the cross-section movement as

$$E_k = \frac{1}{2} \int_0^L \rho A \left(\frac{\partial u}{\partial t}\right)^2 dx.$$

The variational statement of this phenomenon, following Hamilton, is

$$I(u) = \int_{t_1}^{t_2} \frac{A}{2} \int_0^L \left(\rho \left(\frac{\partial u}{\partial t}\right)^2 - E \left(\frac{\partial u}{\partial x}\right)^2\right) dx dt = \text{extremum},$$

with the assumption of constant cross-section. The corresponding Euler-Lagrange differential equation is again comprised of temporal and spatial derivatives and leads to the governing equation

$$E\frac{\partial^2 u}{\partial x^2} = \rho\frac{\partial^2 u}{\partial t^2}.$$

This is formally identical to the one-dimensional wave equation developed in connection with the vibrating string except for the fact that the tension force F is replaced by the modulus of elasticity E here.

The general solution of the differential equation will also be obtained with the same separation process but with different results due to the different boundary conditions. The string was constrained at both ends, but here the beam is fixed on one end and free on the other.

Assuming the same separation

$$u(x,t) = v(t)w(x),$$

and the stationary initial condition, the temporal solution is of the form

$$v(t) = \sin(\omega t),$$

where

$$\omega = \lambda\sqrt{\frac{E}{\rho}}.$$

The spatial solutions are of the form

$$w(x) = d_1\cos(\lambda x) + d_2\sin(\lambda x).$$

The fixed end boundary condition yields

$$w(0) = 0 \rightarrow d_1 = 0,$$

and in order to get non-trivial solution, $d_2 \neq 0$, the velocity at the free end must be zero. This requires $\cos(\lambda L) = 0$, which is satisfied when

$$\lambda L = \frac{\pi}{2}, \frac{3\pi}{2}, \dots$$

hence

$$\lambda_k = \frac{2k-1}{2}\frac{\pi}{L}; k = 1, 2, \dots$$

The natural vibration shapes will be

$$w_k(x) = d_k\sin(\lambda_k x); k = 1, 2, \dots$$

The natural frequencies are computed as

$$\omega_k = \lambda_k \sqrt{\frac{E}{\rho}} = \frac{2k-1}{2}\frac{\pi}{L}\sqrt{\frac{E}{\rho}}, k = 1, 2, \ldots$$

The general fundamental solutions will become

$$u_k(x, t) = d_k \sin(\lambda_k x) \sin(\omega_k t), k = 1, 2, \ldots$$

Using the trigonometric identities of the sin and cos of sums and differences of angles, and ignoring the undefined coefficient, one may derive the form of

$$u_k(x, t) = \frac{1}{2}\left(\cos(\lambda_k x + \omega_k t) + \cos(\lambda_k x - \omega_k t)\right).$$

This solution form is the sum of two traveling waves, originally proposed by d'Alembert. Figure 11.7 demonstrates the traveling wave form of the 2nd vibration shape.

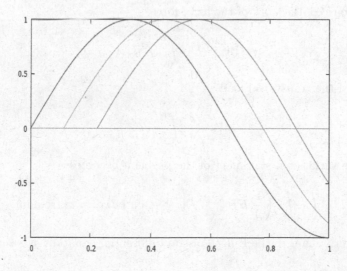

FIGURE 11.7 Axial vibrations of a beam

While the x-axis still represents the longitudinal axis of the beam, the y values in the figure are the axial deformations at that point in the beam. The curves are the second vibration shape in three consecutive time instances, illustrating the wave's propagation from the fixed toward the free end.

11.5   Simultaneous axial and transversal loading of beam

This modeling scenario of the beam is very practical in airplane structures, specifically the components called spars [11]. This is in a sense a combination of the phenomena discussed in two prior sections, except for the fact that the transversal load is allowed to be an external, albeit constant distributed load, not just weight. Nevertheless, the commonality in the variational foundation of the prior sections will be exploited here.

The potential energy captured in the beam is still the internal strain energy as

$$E_s = \frac{1}{2} \int_0^L EI(y'')^2 dx.$$

The work resulting in that deformation is now the superposition of the two external sources, the work of the axial force

$$W_a = F\frac{1}{2} \int_0^L y'^2 dx,$$

and that of the transversal load,

$$W_t = w \int_0^L y dx.$$

Hence, the variational statement of the problem becomes

$$I = \int_{t_1}^{t_2} \int_0^L \left(\frac{1}{2}(EIy''^2 - Fy'^2) - wy \right) dx dt = \text{extremum}.$$

Here we turn again to the Euler-Poisson differential equation, whose components are

$$\frac{\partial f}{\partial y} = -w,$$

$$\frac{d}{dx}\frac{\partial f}{\partial y'} = -\frac{d}{dx}(Fy'),$$

and

$$\frac{d^2}{dx^2}\frac{\partial f}{\partial y''} = \frac{d^2}{dx^2}(EIy'').$$

The Euler-Poisson equation becomes

$$-w + \frac{d}{dx}(Fy') + \frac{d^2}{dx^2}(EIy'') = 0.$$

Introducing again

$$v(x) = y''(x)$$

results in the governing equation

$$\frac{d^2}{dx^2}v(x) + \frac{F}{EI}v(x) = w,$$

or

$$v''(x) + \frac{F}{EI}v(x) = w.$$

The solution of this non-homogeneous differential equation is of the form

$$v(x) = A\sin\left(\sqrt{\frac{F}{EI}}x\right) + B\cos\left(\sqrt{\frac{F}{EI}}x\right) + w\frac{EI}{F}.$$

The boundary conditions at the ends now define different constants than in the prior cases

$$v(0) = 0 \rightarrow 0 + B + w\frac{EI}{F} = 0$$

or

$$B = -w\frac{EI}{F}.$$

The other end produces

$$v(L) = 0 \rightarrow A\sin\left(\sqrt{\frac{F}{EI}}L\right) - w\frac{EI}{F}\cos\left(\sqrt{\frac{F}{EI}}L\right) + w\frac{EI}{F} = 0,$$

from which

$$A = -w\frac{EI}{F}\frac{1 - \cos(\sqrt{\frac{F}{EI}}L)}{\sin(\sqrt{\frac{F}{EI}}L)}.$$

The solution then becomes

$$w(x) = -w\frac{EI}{F}\left(\frac{1 - \cos\left(\sqrt{\frac{F}{EI}}L\right)}{\sin(\sqrt{\frac{F}{EI}}L)}\sin\left(\sqrt{\frac{F}{EI}}x\right) + \cos\left(\sqrt{\frac{F}{EI}}x\right) - 1\right),$$

which is somewhat more difficult to interpret. Clearly when the term

$$\sqrt{\frac{F}{EI}}L = \pi,$$

the denominator is zero and the deformation is infinite. Hence the critical load is still

$$F_{cr} = \frac{\pi^2 EI}{L^2},$$

despite the fact that additional bending loads are present. On the other hand, the actual deformation values are different because the transversal load (that produced the particular solution of the governing equation) also contributed to the deformations.

This is a spectacular manifestation of the superposition principle shared between the engineering and mathematical community. We will leave it at that with the reinforcement of the value of a variational approach to this problem.

11.6 Heat diffusion in a beam

This phenomenon in connection with the beam is of a different physical discipline: heat transfer. Diffusion itself is a general topic applicable to multiple disciplines where some material or energy exchange occurs; however, here we will focus on the transfer of heat. Heat transfer is also multifaceted; dissipation is for example exchanging heat with the environment, while diffusion, the topic of our discussion, is the heat flow inside of a body. Finally, the phenomenon may be transient (changing in time) or steady state when the solution is no longer changing in time, which we will address here in connection with the simple one-dimensional model of the beam.

We assume that the beam is located along the x-axis, it is of length L and insulated on its surface allowing no dissipation of heat. One the other hand, it is connected to the environment at both ends, allowing the control of the boundary temperature there. The beam is of uniform and unit area cross-section, and made of homogeneous material. These are restrictions imposed for simplicity of the discussion, but are not such in practice.

In any diffusion model, there is a density of a certain quantity at a particular location and in time, $u(x, t)$, which is a function assumed to be continuously differentiable in the body. In our case $u(x, t)$ will be the temperature. If there is a change in the density of the quantity, here in temperature, there is a flow accomplishing this change. The speed of this flow is

$$\frac{\partial u}{\partial x},$$

the thermal flow. While we focus on the case when the temperature does not change in time, $\frac{\partial u}{\partial t} = 0$, we will still use partial derivatives to retain generality. Then the flow energy in the body may be written as

$$E_f = \frac{1}{2}k \int_0^L \left(\frac{\partial u}{\partial x}\right)^2 dx,$$

where k is the thermal conductivity coefficient. In most circumstances, there is a heat source of some sort, and the work of this source is

$$W_s = \int_0^L qu(x,t)dx,$$

where q is the generated heat that may also be function of the location, but we will consider it constant here. The flow energy and the work of the source are in balance stated by the functional

$$\int_0^L (E_f - W_s)dx = \int_0^L \left(\frac{1}{2}k \left(\frac{\partial u}{\partial x}\right)^2 - qu\right) dx = \text{extremum.}$$

The Euler-Lagrange differential equation is of the form

$$\frac{\partial f}{\partial u} - \frac{\partial}{\partial x}\frac{\partial f}{\partial u_x} = -q - \frac{\partial}{\partial x}k\frac{\partial u}{\partial x} = 0,$$

from which the steady state heat diffusion governing equation emerges as

$$k\frac{\partial^2 u}{\partial x^2} + q = 0.$$

The general mathematical model is obtained by integrating twice,

$$u(x,t) = -\frac{q}{2k}x^2 + ax + b.$$

The constants of integration may be resolved by boundary conditions. Let us assume that both ends of the beam are kept at constant temperatures. Note that the fact of assuring the temperature stays constant at both ends means that there is a heat extraction or contribution by convection (or maybe radiation) at those ends. These phenomena are not of our concern in this case as they are external to our problem. Their behavior is simply manifested by the temperature boundary conditions.

The boundary conditions of

$$u(0,t) = T_1, u(L,t) = T_2,$$

on the left end will imply

$$b = T_1.$$

On the right end

$$u(L,t) = -\frac{q}{2k}L^2 + aL + T_1 = T_2,$$

from which the second coefficient is resolved as

$$a = \frac{q}{2k}L + \frac{T_2 - T_1}{L}.$$

Finally, the specific mathematical model adhering to the conditions of constant temperatures (T_1, T_2) at both ends and generated heat (q), becomes

$$u(x,t) = -\frac{q}{2k}x^2 + \left(\frac{q}{2k}L + \frac{T_2 - T_1}{L}\right)x + T_1.$$

This is also known as the heat conduction equation. Two sub-cases of this mathematical model are considered. If there is no internal heat generation, $q = 0$, then the solution is simply a linear temperature distribution.

$$u(x,t) = \frac{T_2 - T_1}{L}x + T_1.$$

If there is internal heat generation, but both side temperatures are the same,

$$u(x,t) = -\frac{q}{2k}x^2 + \frac{q}{2k}Lx + T_1,$$

and the solution is a quadratic distribution above the temperatures on the side.

Figure 11.8 shows the temperature profile in the beam along its axis. The physical components, (k, L) were unity for visualization purposes. The top curve is the sub-case when the two side temperatures are the same, $T_1 = T_2$, the second curve is the sub-case with different side temperatures and the third curve, the line represents the case when there is no internal heat generation, $q = 0$.

The above mathematical model may be extended to the transient case, when the change of the temperature still depends on time, hence its temporal derivative also arises. Then the equation becomes

$$k\frac{\partial^2 u}{\partial x^2} + q = \rho c\frac{\partial u}{\partial t}.$$

The new coefficients are the density per unit volume, ρ, and the specific heat, c, of the material of the body. Furthermore, the transient model without a generating source is described by the form

$$k\frac{\partial^2 u}{\partial x^2} = \rho c\frac{\partial u}{\partial t}.$$

This form is most commonly known as the **heat equation** whose algebraic solution follows the techniques used in the wave equation. In those earlier

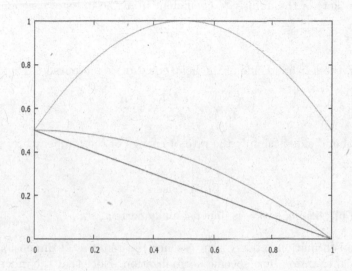

FIGURE 11.8 Temperature distribution in beam

solutions, however, we largely ignored the physics by assigning unit values for physical quantities.

While that approach was useful in producing simplicity in the model generation, the proper way is to transition from a physical to a dimensionless mathematical governing equation. This process will be demonstrated in connection with the heat equation in the next section.

11.6.1 Dimensionless heat equation

The physical components of the heat equation are of different dimensions. Let us introduce the thermal diffusivity coefficient in the form of

$$\kappa = \frac{k}{c\rho},$$

where the right-hand side terms were all defined above. With this, the heat equation is written as

$$\kappa \frac{\partial^2 u}{\partial x^2} = \frac{\partial u}{\partial t}.$$

The first step creating a dimensionless heat equation is to analyze its components using a consistent system of units. Using the SÍ system, the temperature is in Kelvins, time is in seconds and distance in meters. The SI unit (denoted by the bracket) of the diffusivity coefficient from the units of its components becomes

$$[\kappa] = \frac{m^2}{sec}.$$

With this, the left-hand side of the heat equation is measured as

$$\left[\kappa \frac{\partial^2 u}{\partial x^2}\right] = \frac{m^2}{sec} \frac{K}{m^2} = \frac{K}{sec}.$$

The right-hand side is simply the rate of change of the temperature,

$$\left[\frac{\partial u}{\partial t}\right] = \frac{K}{sec},$$

hence the physical equation is dimensionally correct.

The transformation process is in essence the removal of dimensions from the physical equation by a specific normalization. Note that the independent variables, distance (x) and time (t) also have dimensions. Assuming that the beam has physical length L, it is natural to normalize the spatial independent variable as

$$\overline{x} = \frac{x}{L},$$

which produces a dimensionless spatial variable since $[\overline{x}] = 1$. Let us assume a certain time T for normalization purposes, without specifying its value yet. Then

$$\overline{t} = \frac{t}{T},$$

which is also a dimensionless time variable, $[\overline{t}] = 1$. Finally, with a specific temperature U, we obtain

$$\overline{u} = \frac{u}{U},$$

and $[\overline{u}] = 1$. The value of U is usually the maximum temperature feasible for the physical scenario being modeled, somewhat akin to the length normalization in earlier sections.

The derivatives of the heat equation are computed in terms of the dimensionless variables as

$$\frac{\partial u}{\partial t} = \frac{\partial u}{\partial \overline{u}} \frac{\partial \overline{u}}{\partial \overline{t}} \frac{\partial \overline{t}}{\partial t}.$$

Differentiating and substituting result in

$$\frac{\partial u}{\partial t} = \frac{U}{T} \frac{\partial \overline{u}}{\partial \overline{t}}.$$

Similar approach on the spatial derivative brings

$$\frac{\partial u}{\partial x} = \frac{\partial u}{\partial \overline{u}} \frac{\partial \overline{u}}{\partial \overline{x}} \frac{\partial \overline{x}}{\partial x},$$

and

$$\frac{\partial u}{\partial x} = \frac{U}{L} \frac{\partial \overline{u}}{\partial \overline{x}}.$$

Repeating this step yields the second derivative as

$$\frac{\partial^2 u}{\partial x^2} = \frac{U}{L^2} \frac{\partial^2 \overline{u}}{\partial \overline{x}^2}.$$

Finally, substituting into the physical heat equation we obtain

$$\kappa \frac{U}{L^2} \frac{\partial^2 \overline{u}}{\partial \overline{x}^2} = \frac{U}{T} \frac{\partial \overline{u}}{\partial \overline{t}}.$$

We select the normalization time variable as

$$T = \frac{L^2}{\kappa},$$

which is dimensionally correct, $[T] = sec$. Substituting and shortening result in the mathematical, dimensionless heat equation as

$$\frac{\partial^2 \overline{u}}{\partial \overline{x}^2} = \frac{\partial \overline{u}}{\partial \overline{t}}.$$

This is now amenable to purely algebraic solution; however, the results must be reformulated in physical terms. For example, the range of the spatial results is

$$\overline{x} \in (0, 1),$$

from which the physical location is easy to recover by multiplication by L. Similarly, the dimensionless temperature result is simply scaled back by the multiplication by U.

On the other hand, the physical time solution component recovery is more difficult, requiring

$$t = T\overline{t} = \frac{L^2}{\kappa} \overline{t}.$$

While this process was demonstrated in connection with the heat equation, the technique transcends the application areas we discussed. For example, the wave equation may also be rendered dimensionless by using the approach presented herein.

Furthermore, despite the apparent tediousness of the process, in certain real life applications it may be necessary. Many computational mechanics solutions discussed in the next chapter, are executed by software tools devised to be industry independent, and as such, unaware of dimensions and units. It is the users' responsibility to provide the input data in a dimensionless fashion, or at least in a consistent system of units.

Finally, the computational solutions used in large-scale engineering applications apply numerical methods whose accuracy is dependent upon the range of the terms of the variables. Hence the inherent scaling of the process increases the quality of the solution.

12

Computational mechanics

The algebraic difficulties of generating a mathematical model for various phenomena in the last chapter were considerable and may become insurmountable in real-world problems. Computational mechanics is based on the discretization of the geometric continuum and describing its physical behavior in terms of generalized coordinates. Its focus is on computing numerical solutions to practical problems of engineering mechanics.

In order to support the engineering mechanical applications following in the later sections, this chapter starts with their common computational solution technique, the method of finite elements. Computational solutions for elastic bodies, heat conduction and fluid mechanics will be discussed in detail.

12.1   The finite element technique

We introduced the concept of the finite element method as an approximate solution to variational problems in a one-dimensional setting in Section 7.7. The practical importance of the method is in its two- and three-dimensional extension, called the finite element technique, the subject of this section. The Lagrangian formulation used in the next sections for the three-dimensional elasticity, heat conduction and fluid mechanics problems is a generalization of the basis function and discretized approach of the finite element method.

Computational solutions via the finite element technique achieved an unparalleled industrial success. The topic's implementation details cover an extensive territory [13]; hence, we will discuss only the main components here.

These common, application-independent components are:
- automatic discretization of the geometric domain by finite elements,
- computation of the basis functions used in the approximate solutions,
- computation of finite element matrices,
- assembly of the finite element system matrices, and
- solution of the arising linear system or eigenvalue problem.
They are described in detail in the following sections.

12.1.1 Finite element meshing

The discretization of a generic three- or two-dimensional domain is usually by finite elements of simple shapes, such as tetrahedra or triangles. The foundation of many general methods of discretization (commonly called meshing) is the classical Delaunay triangulation method. The Delaunay triangulation technique in turn is based on Voronoi polygons. The Voronoi polygon, assigned to a certain point of a set of points in the plane, contains all the points that are closer to the selected point than to any other point of the set.

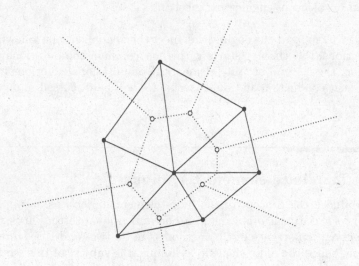

FIGURE 12.1 Delaunay triangularization

In Figure 12.1 the dots represent such a set of points. The irregular (dotted line) hexagon containing one point in the middle is the Voronoi polygon of the point in the center. It is easy to see that the points inside the polygon are closer to the center point than to any other points of the set. It is also quite intuitive that the edges of the Voronoi polygon are the perpendicular bisectors of the line segments connecting the points of the set.

The union of the Voronoi polygons of all the points in the set completely covers the plane. It follows that the Voronoi polygons of two points of the set do not have common interior points; at most they share points on their common boundary.

The definition and process generalize to three dimensions very easily. If the set of points are in space, the points closest to a certain point define a Voronoi polyhedron.

The Delaunay triangulation process is based on the Voronoi polygons by constructing Delaunay edges connecting those points whose Voronoi polygons have a common edge. Constructing all such possible edges will result in the covering of the planar region of our interest with triangular regions, the Delaunay triangles.

The process starts with placing vertices on the boundary of the domain in an equally spaced fashion. The Voronoi polygons of all boundary points are created and interior points are generated gradually proceeding inward by creating Delaunay triangles. This is called the advancing front technique.

The process extends quite naturally and covers the plane as shown in Figure 12.1 with six Delaunay triangles where the dotted lines are the edges of the Voronoi polygons, and the solid lines depict the Delaunay edges. It is known that under the given definitions no two Delaunay edges cross each other.

Finally, in three dimensions, the Delaunay edges are defined as lines connecting points that share a common Voronoi facet (a face of a Voronoi polyhedron). Furthermore, the Delaunay facets are defined by points that share a common Voronoi edge (an edge of a Voronoi polyhedron). In general, each edge is shared by exactly three Voronoi polyhedra; hence, the Delaunay regions' facets are going to be triangles. The Delaunay regions connect points of Voronoi polyhedra that share a common vertex. Since in general the number of such polyhedra is four, the generated Delaunay regions will be tetrahedra. The triangulation method generalized into three dimensions is called tessellation.

12.1.2. Shape functions

We will demonstrate the finite element technique by assuming that the meshed domain in the prior section represents an irregularly shaped membrane problem, and only out of plane deformations of the membrane are considered. This will simplify the presentation of the technique while still capturing its intricacies. The three-dimensional elasticity formulation is simply a generalization of the process presented below.

In the introduction of the finite element method in Chapter 7, we used basis functions to describe the approximate solutions. In order to approximate the solution inside the domain, the finite element technique uses a collection of low order polynomial basis functions. For a triangular element discretization of a two-dimensional domain, as shown in Figure 12.2, bilinear interpolation functions are commonly used in the form:

$$u(x,y) = a + bx + cy.$$

Here u represents any of the q, T or p physical solution quantities introduced in the past three sections.

In order to find the coefficients, let us consider a triangular element in the $x - y$ plane with corner nodes $(x_1, y_1), (x_2, y_2)$ and (x_3, y_3). For this particular triangle we seek three specific basis functions N_i, called shape functions in the finite element field, satisfying

$$N_1 + N_2 + N_3 = 1.$$

We also require that these functions at a certain node point reduce to zero at the other two nodes. This is called the Kronecker property and is presented as

$$N_i = \begin{cases} 1 \text{ at node } i, \\ 0 \text{ at node } \neq i. \end{cases}$$

Furthermore, a shape function is zero along the edge opposite to the particular node at which the shape function is non-zero.

The solution for the nodes of a particular triangular element e can be expressed in matrix form as

$$u_e = \begin{bmatrix} u_1 \\ u_2 \\ u_3 \end{bmatrix} = \begin{bmatrix} 1 & x_1 & y_1 \\ 1 & x_2 & y_2 \\ 1 & x_3 & y_3 \end{bmatrix} \begin{bmatrix} a \\ b \\ c \end{bmatrix}.$$

This system of equations is solved for the unknown coefficients that produce the shape functions

$$\begin{bmatrix} a \\ b \\ c \end{bmatrix} = \begin{bmatrix} 1 & x_1 & y_1 \\ 1 & x_2 & y_2 \\ 1 & x_3 & y_3 \end{bmatrix}^{-1} \begin{bmatrix} u_1 \\ u_2 \\ u_3 \end{bmatrix} = \begin{bmatrix} N_{1,1} & N_{1,2} & N_{1,3} \\ N_{2,1} & N_{2,2} & N_{2,3} \\ N_{3,1} & N_{3,2} & N_{3,3} \end{bmatrix} \begin{bmatrix} u_1 \\ u_2 \\ u_3 \end{bmatrix}.$$

By substituting into the matrix form of the bilinear interpolation function

$$u(x,y) = \begin{bmatrix} 1 & x & y \end{bmatrix} \begin{bmatrix} a \\ b \\ c \end{bmatrix} = \begin{bmatrix} 1 & x & y \end{bmatrix} \begin{bmatrix} N_{1,1} & N_{1,2} & N_{1,3} \\ N_{2,1} & N_{2,2} & N_{2,3} \\ N_{3,1} & N_{3,2} & N_{3,3} \end{bmatrix} \begin{bmatrix} u_1 \\ u_2 \\ u_3 \end{bmatrix},$$

we get

$$u(x, y) = \begin{bmatrix} N_1 & N_2 & N_3 \end{bmatrix} \begin{bmatrix} u_1 \\ u_2 \\ u_3 \end{bmatrix}.$$

Here the N_1, N_2, N_3 shape functions are

$$N_1(x, y) = N_{1,1} + N_{2,1}x + N_{3,1}y,$$
$$N_2(x, y) = N_{1,2} + N_{2,2}x + N_{3,2}y,$$

and

$$N_3(x, y) = N_{1,3} + N_{2,3}x + N_{3,3}y.$$

The shape functions, as their name indicates, solely depend on the coordinates of the corner nodes and the shape of the particular triangular element of the domain. With these we are able to approximate the solution value inside an element in terms of the solutions at the corner node points as

$$u(x, y) = N_1(x, y)u_1 + N_2(x, y)u_2 + N_3(x, y)u_3.$$

The shortcoming of this direct approach is that the coefficients of the shape functions are different for each element and they would have to be computed for all elements in the domain.

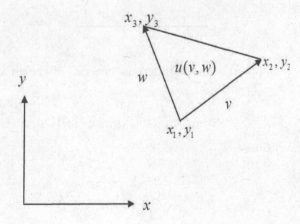

FIGURE 12.2 Parametric coordinates of triangular element

It is practical therefore to generate the shape functions for a standard, parametrically defined element. In that case, the shape functions and their derivatives may be pre-computed and appropriately transformed as was originally proposed in [20]. Let us define a parametric coordinate system for the triangular element as shown in Figure 12.2.

The relationship between the geometric and parametric coordinates is defined by the bidirectional mapping

$$x = x(v, w), y = y(v, w)$$

and

$$v = v(x, y), w = w(x, y).$$

The v axis is directed from node 1 with coordinates (x_1, y_1) to node 2 with coordinates (x_2, y_2). The w axis is directed from node 1 with coordinates (x_1, y_1) to node 3 with coordinates (x_3, y_3). The pairing between the geometric and parametric coordinates of the nodes of the triangle is shown in Table 12.1.

TABLE 12.1
Coordinate pairing of
triangular element

node	x	y	v	w
1	x_1	y_1	0	0
2	x_2	y_2	1	0
3	x_3	y_3	0	1

Let us now compute the shape functions in terms of these parametric coordinates:

$$N_i(v, w) = N_i\left(v(x, y), w(x, y)\right).$$

Specifically, we choose

$$N_1(v, w) = 1 - v - w,$$

$$N_2(v, w) = v,$$

and

$$N_3(v, w) = w.$$

These shape functions also satisfy the Kronecker property stated above and the conditions of polynomial completeness:

$$\sum_{i=1}^{3} N_i(v, w) = 1 - v - w + v + w = 1,$$

$$\sum_{i=1}^{3} N_i(v, w)x_i = x$$

and

$$\sum_{i=1}^{3} N_i(v, w)y_i = y.$$

The last two equations imply that the location of any point inside the element will also be represented by these shape functions as

$$x = N_1(v, w)x_1 + N_2(v, w)x_2 + N_3(v, w)x_3,$$

and

$$y = N_1(v, w)y_1 + N_2(v, w)y_2 + N_3(v, w)y_3.$$

Such elements are called iso-parametric elements since both the geometry and the solution function inside the element are approximated by the same parametric shape functions. Substituting the shape functions we obtain

$$x = (1 - v - w)x_1 + vx_2 + wx_3 = x_1 + (x_2 - x_1)v + (x_3 - x_1)w,$$

and

$$y = (1 - v - w)y_1 + vy_2 + wy_3 = y_1 + (y_2 - y_1)v + (y_3 - y_1)w.$$

This formulation is a crucial component of the standardized element matrix generation as we will see it in the next section.

12.1.3 Element matrix generation

In order to compute a particular matrix for a finite element, we consider all the corner nodes specifying the sides bounding a particular element. For example, the mass of an element is described by the elemental matrix

$$M_e = \rho \int \int_{x,y \in D_e} N(x, y)^T N(x, y) dx dy,$$

where the $N(x, y)$ matrix is the local shape function matrix of the particular element, D_e is its geometric domain and ρ is the density of the material.

Using the parametric coordinates, however, the above elemental mass matrix integral may be evaluated as

$$M_e = \rho \int_{v=0}^{1} \int_{w=0}^{1-v} N(v, w)^T N(v, w) \det \left(\frac{\partial(x, y)}{\partial(v, w)} \right) dw dv.$$

The matrix of the determinant, called the Jacobian matrix, is computed as follows

$$J = \frac{\partial(x,y)}{\partial(v,w)} = \begin{bmatrix} \frac{\partial x}{\partial v} & \frac{\partial x}{\partial w} \\ \frac{\partial y}{\partial v} & \frac{\partial y}{\partial w} \end{bmatrix}.$$

For our triangular element, this is

$$J = \begin{bmatrix} x_2 - x_1 & x_3 - x_1 \\ y_2 - y_1 & y_3 - y_1 \end{bmatrix}$$

and

$$\det(J) = (x_2 - x_1)(y_3 - y_1) - (x_3 - x_1)(y_2 - y_1).$$

This is a different value for each element. However, since it is a constant, it could be moved outside of the integral which will be important. Its value is indicative of the quality of the finite element. A very small value indicates an ill-shaped finite element that will be detrimental to the solution quality as we will see later.

Since the matrix of the shape functions is organized as

$$N(v,w) = \begin{bmatrix} N_1 & N_2 & N_3 \end{bmatrix},$$

the $N(v,w)^T N(v,w)$ product needed for the mass matrix is of the form

$$N(v,w)^T N(v,w) = \begin{bmatrix} N_1^T N_1 & N_1^T N_2 & N_1^T N_3 \\ N_2^T N_1 & N_2^T N_2 & N_2^T N_3 \\ N_3^T N_1 & N_3^T N_2 & N_3^T N_3 \end{bmatrix}.$$

These terms are only functions of the v, w parametric variables; hence, they may be pre-computed as

$$N(v,w)^T N(v,w) = \begin{bmatrix} (1-v-w)^2 & (1-v-w)v & (1-v-w)w \\ v(1-v-w) & v^2 & vw \\ w(1-v-w) & wv & w^2 \end{bmatrix}.$$

The matrix is symmetric and the integral over the parametric domain of the triangular finite element becomes

$$M_e = \rho \det(J) \int_{v=0}^{1} \int_{w=0}^{1-v} N(v,w)^T N(v,w) \, dw \, dv.$$

In this form, the evaluation of the integrals is still cumbersome due to the variable upper limit of the inner integral. They may be further transformed to enable easier integration by the substitution

$$w = \frac{1-v}{2} + \frac{1-v}{2} r,$$

and

$$v = \frac{1}{2} + \frac{1}{2} s.$$

This will of course modify $N(v, w)$, a function of v, w, to $N(s, r)$, a function of s, r and brings the consequences

$$dv = \frac{1}{2}ds$$

and

$$dw = \frac{1 - v}{2}dr.$$

Finally, the integrals become

$$M_e = \rho \det(J) \int_{s=-1}^{1} \frac{1}{2} \int_{r=-1}^{1} N(s,r)^T N(s,r) \left(\frac{1}{4} - \frac{1}{4}s\right) dr ds.$$

These may now be easily integrated for the standard element a priori and only once. During computation of the finite element solution, the standard element matrix is multiplied by values in front of the integrals that are characteristic to the shape of the particular element. This is a fundamental aspect of practical finite element technique.

The generation of the stiffness matrix will require the computation of a matrix containing the derivatives of the shape functions. For our simplified case, the matrix is of the form:

$$B(x, y) = \begin{bmatrix} \frac{\partial N_1}{\partial x} & \frac{\partial N_2}{\partial x} & \frac{\partial N_3}{\partial x} \\ \frac{\partial N_1}{\partial y} & \frac{\partial N_2}{\partial y} & \frac{\partial N_3}{\partial y} \end{bmatrix}.$$

Since the shape functions are defined in terms of the parametric coordinates, the derivatives of the local shape functions are computed by using the chain rule as

$$\frac{\partial N_i}{\partial v} = \frac{\partial N_i}{\partial x}\frac{\partial x}{\partial v} + \frac{\partial N_i}{\partial y}\frac{\partial y}{\partial v}$$

and

$$\frac{\partial N_i}{\partial w} = \frac{\partial N_i}{\partial x}\frac{\partial x}{\partial w} + \frac{\partial N_i}{\partial y}\frac{\partial y}{\partial w}.$$

These relations may be gathered as

$$\begin{bmatrix} \frac{\partial N_i}{\partial v} \\ \frac{\partial N_i}{\partial w} \end{bmatrix} = \begin{bmatrix} \frac{\partial x}{\partial v} & \frac{\partial y}{\partial v} \\ \frac{\partial x}{\partial w} & \frac{\partial y}{\partial w} \end{bmatrix} \begin{bmatrix} \frac{\partial N_i}{\partial x} \\ \frac{\partial N_i}{\partial y} \end{bmatrix}.$$

The first term on the right-hand side is

$$\begin{bmatrix} \frac{\partial x}{\partial v} & \frac{\partial y}{\partial v} \\ \frac{\partial x}{\partial w} & \frac{\partial y}{\partial w} \end{bmatrix} = J^T,$$

as we found it earlier. Hence

$$\begin{bmatrix} \frac{\partial N_i}{\partial v} \\ \frac{\partial N_i}{\partial w} \end{bmatrix} = J^T \begin{bmatrix} \frac{\partial N_i}{\partial x} \\ \frac{\partial N_i}{\partial y} \end{bmatrix}$$

and

$$\begin{bmatrix} \frac{\partial N_i}{\partial x} \\ \frac{\partial N_i}{\partial y} \end{bmatrix} = J^{T,-1} \begin{bmatrix} \frac{\partial N_i}{\partial v} \\ \frac{\partial N_i}{\partial w} \end{bmatrix}.$$

The inverse of the Jacobian matrix may be computed by

$$J^{-1} = \frac{adj(J)}{\det(J)}.$$

This equation clarifies the earlier warning comment about the numerical problems arising from elements with a very small Jacobian determinant that is in the denominator. Hence, we now have arrived at the B matrix with shape function derivatives with respect to the parametric coordinates as

$$B(v,w) = J^{T,-1} \begin{bmatrix} \frac{\partial N_1}{\partial v} & \frac{\partial N_2}{\partial v} & \frac{\partial N_3}{\partial v} \\ \frac{\partial N_1}{\partial w} & \frac{\partial N_2}{\partial w} & \frac{\partial N_3}{\partial w} \end{bmatrix}.$$

Using the terms of the Jacobian matrix we obtained earlier, the adjoint is

$$adj(J^T) = \begin{bmatrix} y_3 - y_1 & x_1 - x_3 \\ y_1 - y_2 & x_2 - x_1 \end{bmatrix},$$

and the determinant becomes

$$\det(J) = (x_2 - x_1)(y_3 - y_1) - (y_2 - y_1)(x_3 - x_1).$$

Therefore, the inverse matrix contains the element specific coordinates. It is easy to find from the preceding that

$$\frac{\partial N_1}{\partial v} = -1, \frac{\partial N_1}{\partial w} = -1,$$

$$\frac{\partial N_2}{\partial v} = 1, \frac{\partial N_2}{\partial w} = 0,$$

and

$$\frac{\partial N_3}{\partial v} = 0, \frac{\partial N_3}{\partial w} = 1.$$

For our specific element, we obtain

$$B(v,w) = J^{T,-1} \begin{bmatrix} -1 & 1 & 0 \\ -1 & 0 & 1 \end{bmatrix}.$$

The elemental stiffness matrix, with the inclusion of the material specific elasticity matrix D, may now be computed as

$$K_e = \det(J) \int_{v=0}^{1} \int_{w=0}^{1-v} B(v,w)^T D B(v,w) dw dv.$$

This element stiffness matrix is of order 3 by 3 for our triangular element with a scalar field solution.

The integral transformation shown in connection with the mass matrix is also executed here as

$$K_e = \det(J) \int_{s=-1}^{1} \frac{1}{2} \int_{r=-1}^{1} B(s,r)^T DB(s,r) \left(\frac{1}{4} - \frac{1}{4}s \right) dr ds.$$

However, due to the content of the B matrix and the presence of the elasticity matrix, this integral cannot be evaluated a priori; it has to be computed during the solution. For the sake of efficiency, the integrals are numerically evaluated, usually by Gaussian quadrature.

The integrals are replaced by weighted sums and the integrand is evaluated at strategically selected points called the Gauss points:

$$K_e = \frac{1}{2} \det(J) \sum_{i=1}^{n} c_i \sum_{j=1}^{n} c_j B^T(s_i, r_j) DB(s_i, r_j) \left(\frac{1}{4} - \frac{1}{4}s_i \right).$$

Table 12.2 shows the $s_i = t_i, r_j = t_j$ Gauss point locations and corresponding c_i weights. Their computation was explained in Section 5.4.1.

TABLE 12.2
Gauss points and weights

n	i	t_i	c_i
1	1	0	2
2	1	−0.577350	1
2	2	0.577350	1
3	1	−0.774597	0.555556
3	2	0	0.888889
3	3	0.774597	0.555556
4	1	−0.861136	0.347855
4	2	−0.339981	0.652146
4	3	0.339981	0.652146
4	4	0.861136	0.347855

For very simple elements, first order ($n = 1$) integration suffices. For elements representing more difficult physics, the second and third order formulae are used. Fourth order integration is sometimes used for quadratic or higher order elements.

Finally, the elemental load vector is also obtained by integrating with the shape function matrix as

$$F_e = \det(J) \int_{v=0}^{1} \int_{w=0}^{1-v} N(v, w)^T \underline{f} \, dw \, dv.$$

Here $\underline{f}$ is the vector of forces acting on the nodes of the element

$$\underline{f} = \begin{bmatrix} f_1 \\ f_2 \\ f_3 \end{bmatrix}.$$

We are now in the position to assemble the elemental matrices and obtain the solution of the problem on the complete domain.

12.1.4   Element matrix assembly and solution

Since the element matrices have been developed in terms of the local (v, w) parametric coordinate system, before assembling they must be transformed to the global (x, y) coordinate system common to all the elements. The coordinates of a point in the two systems are related as

$$\begin{bmatrix} x \\ y \\ 1 \end{bmatrix} = T \begin{bmatrix} v \\ w \\ 1 \end{bmatrix}.$$

The transformation matrix is formed as

$$T = \begin{bmatrix} v_x & w_x & x_1 \\ v_y & w_y & y_1 \\ 0 & 0 & 1 \end{bmatrix},$$

where

$$\underline{v} = v_x \underline{i} + v_y \underline{j}$$

and

$$\underline{w} = w_x \underline{i} + w_y \underline{j}$$

are the vectors in the global system defining the local parametric coordinate axes. The point (x_1, y_1) defines the local element system's origin as was shown in Figure 12.2.

The same transformation is applicable to the solution values. The global solution values are related to the local elemental solution values by the same transformation matrix in the form of

$$\begin{bmatrix} u_{e,x} \\ u_{e,y} \\ 1 \end{bmatrix} = T \begin{bmatrix} u_{e,v} \\ u_{e,w} \\ 1 \end{bmatrix}.$$

Hence, the element solutions in the two systems are related as

$$u_e^g = T u_e$$

or

$$u_e = T^{-1} u_e^g.$$

The u_e^g notation refers to the element solution in the global coordinate system.

Let us now consider an elemental solution with the local element matrix and the local load vector F_e as

$$K_e u_e = F_e.$$

The relationship between the load vector in local terms and its version in the global coordinate system is similar:

$$F_e^g = T F_e,$$

or

$$F_e = T^{-1} F_e^g.$$

Substituting into the elemental solution, we obtain

$$K_e T^{-1} u_e^g = T^{-1} F_e^g,$$

Pre-multiplying by T and exploiting the emerging identity matrix results in

$$T K_e T^{-1} u_e^g = F_e^g,$$

or

$$K_e^g u_e^g = F_e^g.$$

Here

$$K_e^g = T K_e T^{-1}$$

is the element matrix transformed to global coordinates. This transformation follows the element matrix generation and precedes the assembly process.

Finally, the K global stiffness matrix is assembled as

$$K = \sum_{e=1}^{m} L_{ge} K_e^g L_{ge}^T,$$

where L_{ge} is a Boolean matrix mapping the element local node numbers to the global node numbers. If, for example, the element is described by nodes 1, 2 and 3, then the terms in K_e^g contribute to the terms of the 1st, 2nd and 3rd columns and rows of the assembled, global K matrix. Let us assume that another element is adjacent to the edge between nodes 2 and 3 whose third node is numbered 4. The second element's matrix terms will contribute to the 2nd, 3rd and 4th columns and rows of the global matrix.

The individual element matrices are mapped to the global matrix that is of size 4 by 4, reflecting the presence of the 4 node points as

$$L_{ge}K_1^g L_{ge}^T = \begin{bmatrix} K_1(1,1) & K_1(1,2) & K_1(1,3) & 0 \\ K_1(2,1) & K_1(2,2) & K_1(2,3) & 0 \\ K_1(3,1) & K_1(3,2) & K_1(3,3) & 0 \\ 0 & 0 & 0 & 0 \end{bmatrix},$$

and

$$L_{ge}K_2^g L_{ge} = \begin{bmatrix} 0 & 0 & 0 & 0 \\ 0 & K_2(1,1) & K_2(1,2) & K_2(1,3) \\ 0 & K_2(2,1) & K_2(2,2) & K_2(2,3) \\ 0 & K_2(3,1) & K_2(3,2) & K_2(3,3) \end{bmatrix}.$$

Here the subscript is the element index $e = 1, 2$ and the row, column indices in the parenthesis refer to the local element node numbers. The assembled global finite element matrix is then

$$K = \begin{bmatrix} K_1(1,1) & K_1(1,2) & K_1(1,3) & 0 \\ K_1(2,1) & K_1(2,2) + K_2(1,1) & K_1(2,3) + K_2(1,2) & K_2(1,3) \\ K_1(3,1) & K_1(3,2) + K_2(2,1) & K_1(3,3) + K_2(2,2) & K_2(2,3) \\ 0 & K_2(3,1) & K_2(3,2) & K_2(3,3) \end{bmatrix},$$

The assembled global load vector is similarly obtained:

$$F = \sum_{i=1}^2 L_{ge} F_{e,i}^g = \begin{bmatrix} F_1(1) \\ F_1(2) + F_2(1) \\ F_1(3) + F_2(2) \\ F_2(3) \end{bmatrix}.$$

The notation convention is the same as in the element matrix assembly.

The global solution is then obtained from the matrix equation

$$K\underline{u} = F,$$

where K is the global stiffness matrix and F is the global force vector. The global solution vector is

$$\underline{u} = K^{-1}F = \begin{bmatrix} u_1 \\ u_2 \\ u_3 \\ u_4 \end{bmatrix},$$

and the solution inside of a particular element is

$$u(x,y) = N_1^i u_1^i + N_2^i u_2^i + N_3^i u_3^i.$$

The superscript indicates the shape functions and node point values associated with a particular (i-th) element. For the first element in the above hypothetical two element model, $u_j^1 = u_j; j = 1, 2, 3$ and for the second element, $u_j^2 = u_{j+1}; j = 1, 2, 3$.

Naturally the M matrix is similarly transformed and assembled as

$$M = \sum_{e=1}^{m} L_{ge} M_e^g L_{ge}^T.$$

This process is the same for any number of elements contained in the finite element discretization of the geometric model.

12.2 Three-dimensional elasticity

One of the fundamental concepts necessary to understanding continuum mechanical systems is a generic treatment of elasticity described in detail in the classical reference of the subject [17]. When an elastic continuum undergoes a one-dimensional deformation, like in the case of the beam discussed in Section 11.3, Young's modulus was adequate to describe the changes.

For a general three-dimensional elastic continuum we need another coefficient, introduced by Poisson, to capture the three-dimensional elastic behavior. Poisson's ratio measures the contraction of the cross-section, while an object such as a beam is stretched. The ratio ν is defined as the ratio of the relative contraction and the relative elongation:

$$\nu = -\frac{dr}{r} \Big/ \frac{dl}{l}.$$

Here a beam with circular cross-section and radius r is assumed. Poisson's ratio is in the range of zero to $1/2$ and expresses the compressibility of the material. The two constants are also often related as

$$\mu = \frac{E}{2(1+\nu)},$$

and

$$\lambda = \frac{E\nu}{(1+\nu)(1-2\nu)}.$$

Here μ and λ are the so-called Lamé constants.

In a three-dimensional elastic body, the elasticity relations could vary significantly. Let us consider isotropic materials, whose elastic behavior is independent of the material orientation. In this case, Young's modulus is replaced by an elasticity matrix whose terms are only dependent on the Lamé constants as follows

$$D = \begin{bmatrix} \lambda + 2\mu & \lambda & \lambda & 0 & 0 & 0 \\ \lambda & \lambda + 2\mu & \lambda & 0 & 0 & 0 \\ \lambda & \lambda & \lambda + 2\mu & 0 & 0 & 0 \\ 0 & 0 & 0 & \mu & 0 & 0 \\ 0 & 0 & 0 & 0 & \mu & 0 \\ 0 & 0 & 0 & 0 & 0 & \mu \end{bmatrix}.$$

Viewing an infinitesimal cube of the three-dimensional body, there are six stress components on the element,

$$\underline{\sigma} = \begin{bmatrix} \sigma_x \\ \sigma_y \\ \sigma_z \\ \tau_{yz} \\ \tau_{xz} \\ \tau_{xy} \end{bmatrix}.$$

The first three are normal and the second three are shear stresses. There are also six strain components

$$\underline{\epsilon} = \begin{bmatrix} \epsilon_x \\ \epsilon_y \\ \epsilon_z \\ \gamma_{yz} \\ \gamma_{xz} \\ \gamma_{xy} \end{bmatrix}.$$

The first three are extensional strains and the last three are rotational strains. The stress-strain relationship is described by the generalized Hooke's law as

$$\underline{\sigma} = D\underline{\epsilon}.$$

This will be the fundamental component of the computational techniques for elastic bodies. Let us further designate the location of an interior point of the elastic body with

$$\underline{r}(x,y,z) = x\underline{i} + y\underline{j} + z\underline{k} = \begin{bmatrix} x \\ y \\ z \end{bmatrix},$$

and the displacements of the point with

$$\underline{u}(x,y,z) = u\underline{i} + v\underline{j} + w\underline{k} = \begin{bmatrix} u \\ v \\ w \end{bmatrix}.$$

Then the following strain relations hold:

$$\epsilon_x = \frac{\partial u}{\partial x},$$

$$\epsilon_y = \frac{\partial v}{\partial y},$$

and

$$\epsilon_z = \frac{\partial w}{\partial z}.$$

These extensional strains manifest the change of rate of the displacement of an interior point of the elastic continuum with respect to the coordinate directions.

The rotational strains are computed as

$$\gamma_{yz} = \frac{\partial v}{\partial z} + \frac{\partial w}{\partial y},$$

$$\gamma_{xz} = \frac{\partial u}{\partial z} + \frac{\partial w}{\partial x},$$

and

$$\gamma_{xy} = \frac{\partial u}{\partial y} + \frac{\partial v}{\partial x}.$$

These terms define the rate of change of the angle between two lines crossing at the interior point that were perpendicular in the undeformed body and get distorted during the elastic deformation.

The strain energy contained in the three-dimensional elastic continuum is

$$E_s = \frac{1}{2} \int_V \underline{\sigma}^T \underline{\epsilon} \, dV = \frac{1}{2} \int_V \begin{bmatrix} \sigma_x & \sigma_y & \sigma_z & \tau_{yz} & \tau_{xz} & \tau_{xy} \end{bmatrix} \begin{bmatrix} \epsilon_x \\ \epsilon_y \\ \epsilon_z \\ \gamma_{yz} \\ \gamma_{xz} \\ \gamma_{xy} \end{bmatrix} dV.$$

We will also consider distributed forces acting at every point of the volume (like the weight of the beam in Section 11.3), described by

$$\underline{f} = f_x \underline{i} + f_y \underline{j} + f_z \underline{k} = \begin{bmatrix} f_x \\ f_y \\ f_z \end{bmatrix}.$$

The work of these forces is based on the displacements they caused at the certain points and computed as

$$W = \int_V \underline{u}^T \underline{f} dV. = \int_V \begin{bmatrix} u & v & w \end{bmatrix} \begin{bmatrix} f_x \\ f_y \\ f_z \end{bmatrix} dV.$$

In order to evaluate the dynamic behavior of the three-dimensional body, the kinetic energy also needs to be computed. Let the velocities at every point of the volume be described by

$$\underline{\dot{u}}(x, y, z) = \dot{u}\underline{i} + \dot{v}\underline{j} + \dot{w}\underline{k} = \begin{bmatrix} \dot{u} \\ \dot{v} \\ \dot{w} \end{bmatrix}.$$

With a mass density of ρ, assumed to be constant throughout the volume, the kinetic energy of the body is

$$E_k = \frac{1}{2}\rho \int_V \underline{\dot{u}}^T \underline{\dot{u}} dV.$$

We are now in the position to write the variational statement describing the equilibrium of the three-dimensional elastic body:

$$I\left(u(x, y, z)\right) = \int_{t_1}^{t_2} \left(E_k - (E_p - W)\right) dt = \text{extremum},$$

which is of course Hamilton's principle extended with the external work.

The unknown deformation u of the body at every (x, y, z) point is the subject of the computational solution discussed in the next sections.

12.3   Mechanical system analysis

We now consider a mechanical system of a continuum and seek the deformation at any point inside the system. The solution will be obtained by finding an approximate solution of a variational problem in the form of

$$\underline{u}(x, y, z) = \sum_{i=1}^{n} \underline{q}_i N_i(x, y, z).$$

The yet unknown coefficients, the $\underline{q}_i$ values are displacements at $i = 1, 2, \ldots$ discrete locations inside the volume. These are also known as generalized displacements and discussed in an earlier section [14].

The variational statement in detail is

$$I(\underline{u}) = \int_{t_1}^{t_2} \int_V \left(\frac{1}{2} \rho \dot{\underline{u}}^T \dot{\underline{u}} - (\frac{1}{2} \underline{\sigma}^T \underline{\epsilon} - \underline{u}^T \underline{f}) \right) dV dt = \text{extremum}.$$

Let us organize the generalized displacements as

$$\underline{q} = \begin{bmatrix} \underline{q}_1 \\ \cdots \\ \underline{q}_n \end{bmatrix},$$

where, in adherence to our three-dimensional focus

$$\underline{q}_i = \begin{bmatrix} q_{i,x} \\ q_{i,y} \\ q_{i,z} \end{bmatrix}.$$

Using this, the approximate solution becomes

$$\underline{u}(x, y, z) = N \underline{q}$$

with the matrix of basis functions

$$N(x, y, z) = N_x \underline{i} + N_y \underline{y} + N_z \underline{k} = \begin{bmatrix} N_1 & 0 & 0 & \dots & N_n & 0 & 0 \\ 0 & N_1 & 0 & \dots & 0 & N_n & 0 \\ 0 & 0 & N_1 & \dots & 0 & 0 & N_n \end{bmatrix}.$$

The basis functions are usually low order polynomials of x, y, z as shown in Section 12.1.2.

Let us apply this to the terms of our variational problem, starting with the kinetic energy. Assuming that the velocity is also a function of the generalized velocities,

$$\dot{u}(x, y, z) = N \dot{\underline{q}},$$

where

$$\dot{\underline{q}} = \begin{bmatrix} \dot{\underline{q}}_1 \\ \cdots \\ \dot{\underline{q}}_n \end{bmatrix},$$

we obtain

$$E_k = \int_V \frac{1}{2} \rho \dot{\underline{u}}^T \dot{u} dV = \frac{1}{2} \dot{\underline{q}}^t \int_V N^T \rho N dV \dot{\underline{q}}.$$

Introducing the mass matrix

$$M = \int_V N^T \rho N dV,$$

the final form of the kinetic energy becomes

$$E_k = \frac{1}{2}\dot{q}^T M \dot{q}.$$

Now let's focus on the strain energy. Note that the strain is now also expressed in terms of the basis functions. Hence

$$\epsilon(N) = \begin{bmatrix} \sum_{i=1}^{n} q_i^t \frac{\partial N}{\partial x} \\ \sum_{i=1}^{n} q_i^t \frac{\partial N}{\partial y} \\ \sum_{i=1}^{n} q_i^t \frac{\partial N}{\partial z} \\ \sum_{i=1}^{n} q_i^t \left(\frac{\partial N}{\partial z} + \frac{\partial N}{\partial y} \right) \\ \sum_{i=1}^{n} q_i^t \left(\frac{\partial N}{\partial z} + \frac{\partial N}{\partial x} \right) \\ \sum_{i=1}^{n} q_i^t \left(\frac{\partial N}{\partial y} + \frac{\partial N}{\partial x} \right) \end{bmatrix},$$

or in matrix form

$$\underline{\epsilon}(N) = B\underline{q},$$

where the columns of B are

$$B_i = \begin{bmatrix} \frac{\partial N_i}{\partial x} \\ \frac{\partial N_i}{\partial y} \\ \frac{\partial N_i}{\partial z} \\ \frac{\partial N_i}{\partial z} + \frac{\partial N_i}{\partial y} \\ \frac{\partial N_i}{\partial z} + \frac{\partial N_i}{\partial x} \\ \frac{\partial N_i}{\partial y} + \frac{\partial N_i}{\partial x} \end{bmatrix}.$$

With this, the integral becomes

$$\int_V \underline{\epsilon}^T(N) D\underline{\epsilon}(N) dV = \int_V \underline{q}^T B^T DB\underline{q} dV.$$

The total strain energy in the system is

$$E_s = \frac{1}{2}\underline{q}^T \int_V B^T DB dV \underline{q}.$$

Introducing the stiffness matrix of the system as

$$K = \int_V B^T DB dV,$$

the strain energy is of final form

$$E_s = \frac{1}{2}\underline{q}^T K \underline{q}.$$

A similar approach on the external work yields

$$W_e = \int_V \underline{q}^T N^T \underline{f} dV.$$

Introducing the active force vector on the system as

$$F = \int_V N^T \underline{f} dV,$$

this term becomes

$$W_e = \underline{q}^T F.$$

We are ready to find the value of the unknown solution components and will use the extended form of Lagrange's equations of motion

$$\frac{d}{dt} \frac{\partial E_k}{\partial \underline{\dot{q}}} + \frac{\partial E_p}{\partial \underline{q}} = \frac{\partial W_e}{\partial \underline{q}}.$$

The first term is evaluated as

$$\frac{\partial E_k}{\partial \underline{\dot{q}}} = M\underline{\dot{q}}.$$

Then

$$\frac{d}{dt}(M\underline{\dot{q}}) = M\underline{\ddot{q}}.$$

Here the generalized accelerations are

$$\underline{\ddot{q}} = \begin{bmatrix} \ddot{\underline{q}}_1 \\ \cdots \\ \ddot{\underline{q}}_n \end{bmatrix}.$$

The second part results in

$$\frac{\partial E_p}{\partial \underline{q}} = K\underline{q},$$

and the right-hand side brings

$$\frac{\partial W_e}{\partial \underline{q}} = F,$$

The final result is

$$M\underline{\ddot{q}} + K\underline{q} = F.$$

This is the well-known equation of the forced undamped vibration of a three-dimensional elastic body.

12.4   Heat conduction

While staying on the mechanics territory, we now explore the area of heat conduction. This phenomenon occurs when the temperature between two areas of a body differs. In this application, every point in space is associated with a scalar quantity, the temperature, hence these types of problems are called scalar field problems.

For our discussion, we will assume that the body does not deform under the temperature load. This assumption, of course, may be violated in real life. Serious warping of objects left in the sun is a strong example of that scenario.

Two more restrictions we impose. We'll consider two-dimensional problems for simplification of the discussion. We will also only consider the steady state solution case, when the temperature at a certain point is independent of the time. The analytical foundation of this scenario was presented in Section 11.6 in connection with the beam.

The 1D heat conduction problem presented in Section 11.6 may be generalized to two dimensions as [4]

$$k\left(\frac{\partial^2 T}{\partial x^2} + \frac{\partial^2 T}{\partial y^2}\right) + Q = 0,$$

where the temperature and source are now functions of two variables,

$$T = T(x, y), Q = Q(x, y).$$

The k is the thermal conductivity of the material of the object which in general may be a function of the location as well, but considered to be constant here.

This is in essence Poisson's equation in the form of

$$-k\Delta T(x, y) = Q(x, y).$$

Following Section 5.2 where we obtained the variational form of Poisson's equation, the variational form of the heat conduction becomes

$$\int\int_D \left(k\frac{1}{2}\left(\left(\frac{\partial T}{\partial x}\right)^2 + \left(\frac{\partial T}{\partial y}\right)^2\right) - TQ\right) dxdy = \text{extremum}.$$

Following the avenue charted in the last section for the elasticity problem, we will approximate the temperature field in terms of basis functions by

$$T(x, y) = \sum_{i=1}^{n} T_i N_i,$$

where T_i are the temperatures at the discretization points. Then

$$\frac{\partial T(x,y)}{\partial x} = \sum_{i=1}^{n} T_i \frac{\partial N_i}{\partial x},$$

and

$$\frac{\partial T(x,y)}{\partial y} = \sum_{i=1}^{n} T_i \frac{\partial N_i}{\partial y}.$$

Here $T_i = T(x_i, y_i)$ are the temperatures at the discretization locations in the domain. We introduce a vector of these temperatures

$$\underline{T} = \begin{bmatrix} T_1 \\ \dots \\ T_n \end{bmatrix}.$$

We also build a B matrix of the basis function derivatives with columns

$$B_i = \begin{bmatrix} \frac{\partial N_i}{\partial x} \\ \frac{\partial N_i}{\partial y} \end{bmatrix}.$$

Finally we also concatenate the N_i basis functions into the matrix N

$$N = \begin{bmatrix} N_1 \dots N_n \end{bmatrix}.$$

This architecture of the N matrix is simpler than in the case of the elasticity, reflecting the fact that this is a scalar field problem. The elasticity was a vector field problem as the solution quantity at each point was the displacement vector of three dimensions.

Utilizing these matrices we write

$$T(x,y) = N\underline{T},$$

$$\left(\frac{\partial T}{\partial x} \right)^2 + \left(\frac{\partial T}{\partial y} \right)^2 = (B\underline{T})^T B\underline{T},$$

and substituting into the variational problem we obtain

$$\int\int_D \left(k\frac{1}{2}\underline{T}^T B^T B\underline{T} - QN\underline{T} \right) dxdy = \text{extremum}.$$

Introducing a conductivity matrix of

$$K = \int\int_D kB^T B\,dxdy,$$

as well as a source vector of

$$\underline{Q} = \int\int_d QN\,dxdy,$$

we obtain the optimization problem of

$$I(\underline{T}) = \frac{1}{2}\underline{T}^T K \underline{T} - \underline{Q}\underline{T} = \text{extremum}.$$

The solution is obtained from a linear system of equations

$$K\underline{T} = \underline{Q}.$$

This is the computational solution of the two-dimensional heat conduction problem. This two-dimensional process is easily extended to three dimensions with identical computational details.

12.5   Fluid mechanics

As a final application, we discuss a phenomenon when fluid is partially or fully surrounded by an external structure and the dissipation of energy into the surrounding space is negligible [19].

Assuming small motions, the equilibrium of a compressible fluid inside a cavity is governed by the Euler equation derived in Section 10.5

$$\rho\ddot{u} = -\nabla p,$$

where $\ddot{u}$ is the acceleration of the particles and p is the pressure in the fluid. Furthermore, ρ is the density and ∇ is the differential operator. The explicit vector notation is again omitted for the physical vectors, but the linear algebraic vectors are marked as such.

We also assume locally linear pressure-velocity behavior of the fluid as

$$p = -b\nabla u,$$

where b is the so-called bulk modulus related to the density of the fluid and the speed of sound. Differentiating twice with respect to time and substituting the Euler equation, we get Helmholtz's equation describing the behavior of the fluid:

$$\frac{1}{b}\ddot{p} = \nabla\left(\frac{1}{\rho}\nabla p\right).$$

The following boundary conditions are also applied. At a structure-fluid interface

$$\frac{\partial p}{\partial \underline{n}} = -\rho\ddot{u}_n, \tag{12.1}$$

where $\underline{n}$ is the direction of the outward normal. At free surfaces:

$$u = p = 0.$$

Since the equilibrium differential equation of the physical phenomenon is given at this time, the inverse problem approach introduced in Chapter 5 will be used again. Accordingly, for the differential equation of

$$Au = 0,$$

the variational problem of

$$I(u) = (Au, u)$$

applies. Here the inner product is defined over the continuum. For our case, this results in

$$\int \int \int_V \left(\frac{1}{b} \ddot{p} - \frac{1}{\rho} \nabla \cdot \nabla p \right) p \, dV = 0. \tag{12.2}$$

Following the earlier sections, we will also assume that the pressure field is approximated by basis functions as:

$$p(x, y, z) = \sum_{i=1}^{n} N_i p_i = N\underline{p}.$$

The same holds for the derivatives:

$$\ddot{p}(x, y, z) = N\underline{\ddot{p}}.$$

Separating the two parts of Equation (12.2), the first yields

$$\int_V \frac{1}{b} \ddot{p} p \, dV = \int_V \frac{1}{b} p \ddot{p} \, dV = \underline{p}^T \int_V \frac{1}{b} N^T N \, dV \underline{\ddot{p}}.$$

Introducing the mass matrix

$$M = \int_V \frac{1}{b} N^T N \, dV,$$

this term simplifies to

$$\int_V \frac{1}{b} \ddot{p} p \, dV = \underline{p}^T M \underline{\ddot{p}}.$$

Let us now turn our attention to the second part of Equation (12.2). Integrating by parts yields

$$-\int_V \left(\frac{1}{\rho} \nabla \cdot \nabla p \, p \right) dV = \int_V \frac{1}{\rho} \nabla p \cdot \nabla p \, dV - \int_S \frac{1}{\rho} \nabla p \, p \, dS.$$

From the above assumptions, it follows that

$$\nabla p = \nabla N \underline{p},$$

and we obtain

$$\underline{p}^T \int_V \left(\frac{1}{\rho} \nabla N^T \right) \nabla N dV \, \underline{p} + \underline{p}^T \int_S N^T \ddot{u}_n dS.$$

Here the boundary condition stated in Equation (12.1) was used. Introducing

$$K = \int_V \frac{1}{\rho} \nabla N^T \nabla N dV,$$

the first part simplifies to

$$\underline{p}^T K \underline{p}.$$

The force exerted on the boundary by the surrounding structure is

$$F = \int_S N^T \ddot{u}_n dS.$$

Substituting and reordering yields

$$\underline{p}^T M \ddot{\underline{p}} + \underline{p}^T K \underline{p} + \underline{p}^T F = 0.$$

Finally, the equilibrium equation is

$$M \ddot{\underline{p}} + K \underline{p} + F = 0.$$

This, as all the similar problems of this chapter, may be solved by efficient numerical linear algebra computations and will not be discussed further here.

In conclusion, let us emphasize the fact that in all three mechanical engineering disciplines (structural elasticity, heat conduction and fluid mechanics) we used the same computational technique to model the behavior of the physical phenomenon over the geometric domain.

Furthermore, it is important to notice the finite element technique's transcendence of the multiple engineering disciplines. For example, as demonstrated in [3], the governing equation in electrostatics is also Poisson's equation, albeit the participating terms have different physical meaning.

The applicability of a certain variational problem to unrelated disciplines is straightforward; one only needs to adhere to the differences in the physics. This fact makes the techniques demonstrated in this book extremely useful in modeling of various phenomena in diverse engineering disciplines.

Solutions to selected exercises

Section 1.6

1.
$y = \frac{1}{2}x^2 + c_1 x + c_2.$

2.
$y = -x^2 + c_1 x + c_2.$

3.
$y = c_1 x + c_2.$

4.
$y = c_1 \ln |x| + \frac{1}{2}x + c_2.$

5.
$y = 2 - \sqrt{5 - x^2}.$

6.
$y = \sin(4x).$

7.
$y = \frac{\ln |1+x|}{\ln 2}.$

8.
$y = \frac{c_1 x^4}{8} + c_2$

9.
$y = c_1 e^x + c_2 e^{-x} + \frac{1}{2}x.$

10.
$y = c_1 + c_2 x + \frac{1}{2}x^2.$

Section 2.5

1.
$y = \sqrt{8x - x^2}.$

2.
$y = \pm 2\cos(k\pi x), k = 1, 2, 3, ..$

3.
$y = \pm\sqrt{\frac{2}{\pi}}\cos(kx), k = 1, 2, 3, ..$

4.
$y = c_1\cos(x) + \frac{2}{pi}.$

5.
$(x - c_1)^2 + (y - c_2)^2 = \lambda^2$

6.
$y = \lambda\cosh(\frac{x - c_1}{c_2}).$

7.
$y = \cos(x) + 2\sin(x) - 1.$

8.
$y = \frac{1}{12}x^3 + \frac{21}{8}x^2 - \frac{5}{8}x - \frac{1}{12}.$

9.
$y = (-2 \pm x_1\sqrt{5})x + c_2.$

10.
$y = -18x^2 + 20x.$

Section 3.6

1.
$2xu_x + x^2 u_{xx} + 2yu_y + y^2 u_{yy} = 0.$

2.
$u_{tt} - c^2 u_{xx} = 0.$

3.
$u_{xx} + u_{yy} = 0.$

4.
$\Delta u = \lambda u; u = u(x, y, z)$

5.
$1 - z_{xx} - z_{yy} = 0.$

6.
$\ddot{y} = y^2 + 2xy; \ddot{x} = -x^2 - 2xy;$

7.
$x = c_1 e^t + c_2 e^{-t}; y = c_3 e^t + c_4 e^{-t}.$

8.
$x = c_1 e^t + c_2 e^{-t}; y = c_3 e^t + c_4 e^{-t}; z = c_5 e^t + c_6 e^{-t}.$

Section 4.5

6.
$y(x) = a e^{\sqrt{2}x} + b e^{-\sqrt{2}x} + c\cos(\sqrt{2}x) + d\sin(\sqrt{2}x).$

Section 5.4

1. $\frac{d}{dx}\left[x\frac{dy}{dx}\right] + (-\frac{1}{x} + \lambda x)y = 0.$
$x^2 y'' + xy' + (x^2 - \lambda)y = 0, \lambda = n^2.$

2. $\frac{d}{dx}\left[\sqrt{1-x^2}\frac{dy}{dx}\right] + \frac{\lambda}{\sqrt{1-x^2}}y = 0.$
$(1 - x^2)y'' - xy' + \lambda y = 0, \lambda = n^2.$

3. $\frac{d}{dx}\left[e^{-\frac{x^2}{2}}\frac{dy}{dx}\right] + \lambda e^{-\frac{x^2}{2}}y = 0.$
$y'' - xy' + \lambda y = 0, \lambda = 2n.$

4. $\frac{d}{dx}\left[xe^{-x}\frac{dy}{dx}\right] + \lambda e^{-x}y = 0.$
$xy'' + (1 - x)y' + \lambda y = 0, \lambda = n.$

5. $\frac{d}{dx}\left[(1-x^2)\frac{dy}{dx}\right] + \lambda y = 0.$

$(1-x^2)y'' - 2xy' + \lambda y = 0, \lambda = n(n+1).$

They are the Bessel, Chebyshev, Hermite, Laguerre and Legendre equations, in order.

Section 6.6

1. $u(x,y,t) = ce^{-\lambda t}e^{\pm kx}e^{\pm\sqrt{\lambda^2-k^2}y}.$

2. $u(x,y) = 2\sqrt{a_1 x} + \sqrt{2-a_1}y + a_2.$

4. $y_2(x) = 1 - \frac{5}{14}(2x - x^2); y_1 = 1.$

5. $y_2(x) = -\frac{21}{25}(-x + \frac{x^3}{3}); y_1 = 0.$

Section 7.8

1. $y = \frac{1}{2}(x + x^2).$

2. $y = \frac{5}{4}x - \frac{1}{4}x^2.$

3. $y = 2x - x^2.$

4. $y = x^3.$

5. $y = x^3.$

Notations

Notation	Meaning
$f(x)$	Function of one variable
$f(x, y), F(x, y)$	Function of two variables
r	Radius of curvature
g	Acceleration of gravity
$\underline{r}$	Vector in Cartesian coordinates
$\underline{r}_x$	First partial derivative with respect to x
$\underline{r}_y$	First partial derivative with respect to y
$\underline{\dot{r}}$	First parametric derivative
$\underline{\ddot{r}}$	Second parametric derivative
p	Pressure
s	Arc length
$s(t)$	One-dimensional spline function
$s(u, v)$	Two-dimensional spline function
$y'(x), f'(x)$	First derivative
$y''(x), f''(x)$	Second derivative
$\underline{n}$	Normal vector
$\underline{t}$	Tangent vector
$\underline{b}$	Bi-normal vector
∇	Gradient operator
Δ	Laplace operator
κ	Curvature
κ_g	Geodesic curvature
κ_n	Normal curvature
κ_m	Mean curvature
δI	Variation of integral functional
Γ_{ij}^k	Christoffel symbols
σ	Stress
ϵ	Strain
ν	Poisson's ratio
λ	Eigenvalue
ρ	Material density

$[B]$	Shape function derivative matrix
$B_{i,k}(t)$	B-spline basis function
E, F, G	First fundamental quantities
E	Young's modulus
$[D]$	Elasticity matrix
F	Active force
F_{cr}	Critical buckling force
G	Green's function
E_s	Strain energy
E_k	Kinetic energy
E_p	Potential energy
$I()$	Integral functional
I	Moment of inertia
$[J]$	Jacobian matrix
$L_{1...5}$	Lagrange points
$[K]$	Stiffness matrix
$[M]$	Mass matrix
M	Momentum
$[N]$	Shape function matrix
Q	Heat source
S	Surface area
T	Surface (traction) force
$[T]$	Temperature matrix
V	Volume
W_e	External work

References

[1] Barnhill, R. E. and Riesenfeld, R. F.: Computer-aided geometric design, Academic Press, New York, 1974

[2] Beyer, W. H.: Standard mathematical tables, 25th ed., CRC Press, Boca Raton, Florida, 1978

[3] Brauer, J. R.: Magnetic actuators and sensors, Wiley, New Jersey, 2006

[4] Carslaw, H. S. and Jaeger, J. C.: Conduction of heat in solids, Oxford University Press, Cambridge, 1959

[5] Forsyth, A. R.: Lectures on differential geometry of curves and surfaces, Cambridge University Press, 1950

[6] Fox, C.: An introduction to calculus of variations, Oxford University Press, New York, 1950

[7] Gelfand, I. M. and Fomin, S. V.: Calculus of variations, Dover Publications, New York, 1963

[8] Gould, S. H.: Variational methods for eigenvalue problems, University of Toronto Press, Toronto, 1957

[9] Hilbert, D. -Courant, R.: Methods of mathematical physics, Springer, New York, 1953

[10] Kantorovich, L. V. and Krylov, V. I.: Approximate methods of higher analysis, Inter-science, New York, 1958

[11] Karman, T. V. and Biot, M. A.: Mathematical methods in engineering, McGraw-Hill, New York, 1940

[12] Komzsik, L.: Approximation techniques for engineers, 2nd ed, CRC Press, Taylor & Francis Group, Boca Raton, Florida, 2017

[13] Komzsik, L.: Computational techniques of finite element analysis, 2nd ed, CRC Press, Taylor & Francis Group, Boca Raton, Florida, 2011

[14] Lanczos, C.: The variational principles of mechanics, University of Toronto Press, Toronto, 1949

[15] Sommerfeld, A.: Partial differential equations in physics, Academic Press, New York, 1967

[16] Struwe, M.: Variational Methods, Springer, Berlin, 2008

[17] Timoshenko, S. and Goodier, J. N.: Theory of elasticity, McGraw-Hill, New York, 1951

[18] Weinstock, R.: Calculus of variations, McGraw-Hill, New York, 1953

[19] Warsi, Z. U. A.: Fluid dynamics, CRC Press, Taylor & Francis Group, Boca Raton, Florida, 2006

[20] Zienkiewicz, O. C.: The finite element method, McGraw-Hill, New York, 1968

Index

Printed in the United States
by Baker & Taylor Publisher Services